SpringerBriefs in Electrical and Computer Engineering

Computational Electromagnetics

Series editor

Rakesh Mohan Jha, Bangalore, India

More information about this series at http://www.springer.com/series/13885

Balamati Choudhury · Aniruddha R. Sonde
Rakesh Mohan Jha

Terahertz Antenna Technology for Space Applications

 Springer

Balamati Choudhury
Centre for Electromagnetics
CSIR-National Aerospace Laboratories
Bangalore, Karnataka
India

Rakesh Mohan Jha
Centre for Electromagnetics
CSIR-National Aerospace Laboratories
Bangalore, Karnataka
India

Aniruddha R. Sonde
Centre for Electromagnetics
CSIR-National Aerospace Laboratories
Bangalore, Karnataka
India

ISSN 2191-8112   ISSN 2191-8120   (electronic)
SpringerBriefs in Electrical and Computer Engineering
ISSN 2365-6239   ISSN 2365-6247   (electronic)
SpringerBriefs in Computational Electromagnetics
ISBN 978-981-287-798-7   ISBN 978-981-287-799-4   (eBook)
DOI 10.1007/978-981-287-799-4

Library of Congress Control Number: 2015947808

Springer Singapore Heidelberg New York Dordrecht London

Printed on acid-free paper

Springer Science+Business Media Singapore Pte Ltd. is part of Springer Science+Business Media
(www.springer.com)

To Dr. Sudhakar K. Rao

In Memory of Dr. Rakesh Mohan Jha
Great scientist, mentor, and excellent
human being

Dr. Rakesh Mohan Jha was a brilliant contributor to science, a wonderful human being, and a great mentor and friend to all of us associated with this book. With a heavy heart we mourn his sudden and untimely demise and dedicate this book to his memory.

Foreword

National Aerospace Laboratories (NAL), a constituent of the Council of Scientific and Industrial Research (CSIR), is the only civilian aerospace R&D Institution in India. CSIR-NAL is a high-technology institution focusing on various disciplines in aerospace and has a mandate to develop aerospace technologies with strong science content, design and build small and medium-size civil aircraft prototypes, and support all national aerospace programs. It has many advanced test facilities including trisonic wind tunnels which are recognized as national facilities. The areas of expertise and competencies include computational fluid dynamics, experimental aerodynamics, electromagnetics, flight mechanics and control, turbo-machinery and combustion, composites for airframes, avionics, aerospace materials, structural design, analysis, and testing. CSIR-NAL is located in Bangalore, India, with the CSIR Headquarters being located in New Delhi.

CSIR-NAL and Springer have recently signed a cooperation agreement for the publication of selected works of authors from CSIR-NAL as Springer book volumes. Within these books, recent research in the different fields of aerospace that demonstrate CSIR-NAL's outstanding research competencies and capabilities to the global scientific community will be documented.

The first set of five books are from selected works carried out at the CSIR-NAL's Centre for Electromagnetics and are presented as a part of the series SpringerBriefs in Computational Electromagnetics, which is a sub-series of SpringerBriefs in Electrical and Computer Engineering.

CSIR-NAL's Centre for Electromagnetics mainly addresses issues related to electromagnetic (EM) design and analysis carried out in the context of aerospace engineering in the presence of large airframe structures, which is vastly different and in contrast to classical electromagnetics and which often assumes a free-space ambience. The pioneering work done by the Centre for Electromagnetics in some of these niche areas has led to founding the basis of contemporary theories. For example, the geodesic constant method (GCM) proposed by the scientists of the Centre for Electromagnetics is immensely popular with the peers worldwide, and forms the basis for modern conformal antenna array theory.

The activities of the Centre for Electromagnetics consist of (i) Surface modeling and ray tracing, (ii) Airborne antenna analysis and siting (for aircraft, satellites and SLV), (iii) Radar cross section (RCS) studies of aerospace vehicles, including radar absorbing materials (RAM) and structures (RAS), RCS reduction and active RCS reduction, (iv) Phased antenna arrays, conformal arrays, and conformal adaptive array design, (v) Frequency-selective surface (FSS), (vi) Airborne and ground-based radomes, (vii) Metamaterials for aerospace applications including in the Terahertz (THz) domain, and (viii) EM characterization of materials.

It is hoped that this dissemination of information through these SpringerBriefs will encourage new research as well as forge new partnerships with academic and research organizations worldwide.

Shyam Chetty
Director
CSIR-National Aerospace Laboratories
Bangalore, India

Preface

In the pursuit of extracting more information about our universe, terahertz technology is taking center stage in achieving high data rate communications. This technology uses the spectrum that lies in between the microwave and the infrared region with frequencies ranging from 300 GHz to 3 THz. The terahertz band provides a transition between the electronic and the photonic regions thus adopting important characteristics from these regimes. These characteristics corresponding with the progress in semiconductor technology help researchers to exploit this hitherto unexplored domain for a variety of applications.

This relatively new technology is now being extensively researched as it finds numerous applications across disciplines such as communication systems, bio-medical imaging, and security systems, to name a few. This diversification has forced the researchers to come up with feasible and novel devices for the sources, antennas, detectors, etc. The two major challenges where technology needs further exploration, are the implementation of compact, consistent and cheap terahertz sources, and high sensitivity terahertz detectors. Additional complexities involved in scaling from the microwave domain also need to be specifically mapped and managed.

The lack of feasible terahertz sources has hindered the direct applications of terahertz technology in various areas. Realization of nanotechnology has led to the creation of devices such as QCLs, uni-travelling-carrier photo diodes and RTDs providing new avenues to generate terahertz radiation. Critical improvements in the design and characteristics of antennas and antenna arrays are for realizing the desired requirements. Scaling of the existing technologies to adapt to the terahertz region leads to inefficient designs and does not give optimum solutions. Advancements in micromachining techniques have broken the traditional norms of manufacturing, allowing one to fabricate novel, highly integrated devices. Integration of the functional elements provides better transmission, radiation with a higher order of reliability and compactness. Fabrication of arrays to improve the directivity and the gain of the systems has become much more convenient.

The advent of PBG structures and metamaterials has helped to optimize the existing technologies to give improved directivities and gains. Adopting new materials and nanostructures such as graphene has further miniaturized antenna technology while maintaining the desired output levels. Terahertz antenna characterization of bandwidth, impedance, polarization, etc., has not yet been methodically structured, and it continues to be a major research challenge. Developing accurate models to characterize and simulate the designed antennas and state-of-the-art compact antenna test ranges to measure and test the antennas is paramount.

Space applications incorporate all the components of the terahertz technology, viz., communications, detection, and imaging. Concentrations of the order of parts per billion are encountered in astronomical observations, and thus there is a need to improve the sensitivity of the detectors and the receivers. Communication signals decrease with the square of the distance between the receiver and the transmitter. The distances encountered in deep-space communications are very large compared to the distances between satellites and ground stations. High data rates can be achieved using terahertz communications. The basis of all these components is the antennas that transmit and receive these terahertz signals. This brief describes an overview of this rapidly developing technology.

Balamati Choudhury
Aniruddha R. Sonde
Rakesh Mohan Jha

Acknowledgments

We would like to thank Mr. Shyam Chetty, Director, CSIR-National Aerospace Laboratories, Bangalore for his permission and support to write this SpringerBrief.

We would also like to acknowledge valuable suggestions from our colleagues at the Centre for Electromagnetics, Dr. R.U. Nair, Dr. Hema Singh, Dr. Shiv Narayan, and Mr. K.S. Venu and their invaluable support during the course of writing this book.

But for the concerted support and encouragement from Springer, especially the efforts of Suvira Srivastav, Associate Director, and Swati Mehershi, Senior Editor, Applied Sciences & Engineering, it would not have been possible to bring out this book within such a short span of time. We very much appreciate the continued support by Ms. Kamiya Khatter and Ms. Aparajita Singh of Springer towards bringing out this brief.

Balamati Choudhury
Aniruddha R. Sonde
Rakesh Mohan Jha

Contents

Terahertz Antenna Technology for Space Applications   1
1   Introduction   1
      1.1   Characteristics and Advantages   2
      1.2   Terahertz Sources.   3
2   Applications   4
      2.1   Communications   5
      2.2   Detection and Imaging   8
3   Terahertz Antennas   10
      3.1   Photoconductive Antennas.   10
      3.2   MEMS-Based Antennas   14
      3.3   Photonic Band Gap-Based Antennas.   14
      3.4   Arrays   17
      3.5   Nanostructures.   18
      3.6   Graphene-Based Antennas.   19
      3.7   Leaky Wave Antennas Using Graphene-Based
           High Impedance Surfaces   21
      3.8   Substrate Integrated Antennas   21
      3.9   THz Antennas with Artificial Dielectric Superstrate   21
4   Antenna Measurement and Testing.   22
      4.1   Characterization   22
      4.2   Terahertz Antenna Testing.   24
5   Space Applications.   26
      5.1   Deep Space Network   29
6   Conclusions   31
References   31

Appendix A: Prefixes   35

Appendix B: Physical Constants   37

Appendix C: Maxwell's Equations   39

Appendix D: Antenna Properties   41

About the Book   45

Author Index   47

Subject Index   49

About the Authors

Dr. Balamati Choudhury is currently working as a scientist at Centre for Electromagnetics of CSIR-National Aerospace Laboratories, Bangalore, India since April 2008. She obtained her M.Tech. (ECE) degree in 2006 and Ph.D. (Engg.) degree in Microwave Engineering from Biju Patnaik University of Technology (BPUT), Rourkela, Orissa, India in 2013. During the period of 2006–2008, she was a Senior Lecturer in Department of Electronics and Communication at NIST, Orissa India. Her active areas of research interests are in the domain of soft computing techniques in electromagnetics, computational electromagnetics for aerospace applications and metamaterial design applications. She was also the recipient of the CSIR-NAL Young Scientist Award for the year 2013–2014 for her contribution to the area of Computational Electromagnetics for Aerospace Applications. She has authored and co-authored over 100 scientific research papers and technical reports including a book and three book chapters. Dr. Balamati is also an Assistant Professor of AcSIR, New Delhi.

Mr. Aniruddha Ramakrishna Sonde interned as a visiting student at the Centre for Electromagnetics, CSIR-National Aerospace Laboratories (CSIR-NAL), and worked on terahertz communication, deep space networks and terahertz antennas.

Dr. Rakesh Mohan Jha was Chief Scientist & Head, Centre for Electromagnetics, CSIR-National Aerospace Laboratories, Bangalore. Dr. Jha obtained a dual degree in BE (Hons.) EEE and M.Sc. (Hons.) Physics from BITS, Pilani (Raj.) India, in 1982. He obtained his Ph.D. (Engg.) degree from Department of Aerospace Engineering of Indian Institute of Science, Bangalore in 1989, in the area of computational electromagnetics for aerospace applications. Dr. Jha was a SERC (UK) Visiting Post-Doctoral Research Fellow at University of Oxford, Department of Engineering Science in 1991. He worked as an Alexander von Humboldt Fellow at the Institute for High-Frequency Techniques and Electronics of the University of Karlsruhe, Germany (1992–1993, 1997). He was awarded the Sir C.V. Raman Award for Aerospace Engineering for the Year 1999. Dr. Jha was elected Fellow of INAE in 2010, for his contributions to the EM Applications to Aerospace Engineering. He was also the Fellow of IETE and Distinguished Fellow of ICCES.

Dr. Jha has authored or co-authored several books, and more than five hundred scientific research papers and technical reports. He passed away during the production of this book of a cardiac arrest.

Abbreviations

AUT	Antenna Under Test
BWO	Backward Wave Oscillator
CATRs	Compact Antenna Test Ranges
CFRP	Carbon Fibre Reinforced Polymers
CNT	Carbon Nanotube
COBE	Cosmic Microwave Background Experiment
EBG	Electromagnetic Band Gap
FEM	Finite Element Method
HAD	Hybrid Deployable Antenna
HEB	Hot Electron Bolometer
HEMT	High Electron Mobility Transistor
ITO	Indium Tin Oxide
ITU	International Telecommunication Union
LT-GaAs	Low Temperature Gallium Arsenide
MEMS	MicroElectronics Mechanical Systems
MMIC	Monolithic Microwave Integrated Circuits
OSI	Open Systems Interconnection
PBG	Photonic Band Gap
QCL	Quantum Cascade Laser
RMS	Root Mean Square
ROC	Radiometer on Chip
RTD	Resonant Tunnelling Diode
SIS	Super Conductor–Insulator–Super Conductor
SNR	Signal-to-Noise-Ratio
SWAS	Submillimeter Wave Astronomy Satellite
TDS	Time-Domain Spectroscopy
VLSI	Very Large Scale Integration
WLAN	Wireless Local Area Network

List of Figures

Figure 1   The electromagnetic spectrum around the THz
region 2

Figure 2   Secure military communication applications 5

Figure 3   Wireless personal area network 5

Figure 4   Block diagram of terahertz-TDS spectroscopy
system 9

Figure 5   Block diagram of terahertz frequency spectroscopy
system 10

Figure 6   Circuit diagram for a conventional antenna 11

Figure 7   Circuit diagram for a photoconductive antenna 11

Figure 8   Schematic of a photoconductive antenna 12

Figure 9   A schematic of a photoconductive horn antenna 13

Figure 10   Threewire folded dipole antenna 13

Figure 11   Schematic of a PBG antenna 15

Figure 12   Schematic of a PBG structure 16

Figure 13   Schematic of a high gain PBG-based microstrip
antenna 17

Figure 14   Return loss characteristics of the PBG-based microstrip
antenna 18

Figure 15   Radiation characteristics of the PBG-based microstrip
antenna 19

Figure 16   Graphene-based reflector array 20

Figure 17   Graphene-based reflector array 20

Figure 18   Conventional far field measurements 24

Figure 19   Compact antenna test range for far field measurements
(Reflector based) 25

Figure 20   Compact antenna test range for far field measurements
(Lens based) 25

Figure 21   Compact antenna test range for far field measurements
            (Hologram based).   26
Figure 22   Side view of an amplitude hologram.   26
Figure 23   Terahertz instruments corresponding to frequency
            of operation and sensitivity   28
Figure 24   Terahertz heterodyne receiver schematic   28
Figure 25   Indirect links in a deep space network   30

List of Tables

Table 1   The terahertz domain in different units.   3

Table 2   Terahertz applications timeline   4

Table 3   Minimum distances for the interfering signal to be negated
at the radio telescope.   7

Table 4   Important differences between millimeter and submillimeter
antenna technology   11

Table 5   Type of waveguide and corresponding attenuations   16

Table 6   Terahertz instruments for space developed by
various agencies   27

Terahertz Antenna Technology for Space Applications

Abstract The terahertz (THz) band provides a transition between the electronic and the photonic regions thus adopting important characteristics from these regimes. These characteristics corresponding with the progress in semiconductor technology has enabled researchers to exploit hitherto unexplored domains including satellite communication, biomedical imaging, security systems, etc. This book, explores the terahertz antenna technology toward implementation of compact, consistent, and cheap terahertz sources, as well as the high-sensitivity terahertz detectors. The advances in new materials and nanostructures such as graphene will be helpful in miniaturization of antenna technology while simultaneously maintaining the desired output levels. Terahertz antenna characterization of bandwidth, impedance, polarization, etc. has not yet been methodically structured and it continues to be a major research challenge. This book addresses these issues besides including the advances of terahertz technology in space applications worldwide, along with possibilities of using this technology in deep space networks.

Keywords Terahertz antenna · Terahertz sources · Terahertz detectors · Deep space network · Aerospace applications

1 Introduction

Terahertz radiation is the spectrum of light in the 300 GHz to 3 THz (10^{12}) range. It is also referred to as the submillimeter band because of the corresponding wavelengths $(\lambda = c/f)$ of 1–0.1 mm respectively. The band, as seen in Fig. 1, lies in between the microwave and the infrared region. This results in a number of characteristics and applications that can be adopted from these established domains by this relatively unexplored band (for example, steering a beam in a particular direction, in the microwave region, a phased array antenna is used, whereas optical components are used in the infrared region. But for the terahertz region both options are available). Unlike the microwave region whose concepts were developed and

© The Author(s) 2016   1
B. Choudhury et al., *Terahertz Antenna Technology for Space Applications*,
SpringerBriefs in Computational Electromagnetics,
DOI 10.1007/978-981-287-799-4_1

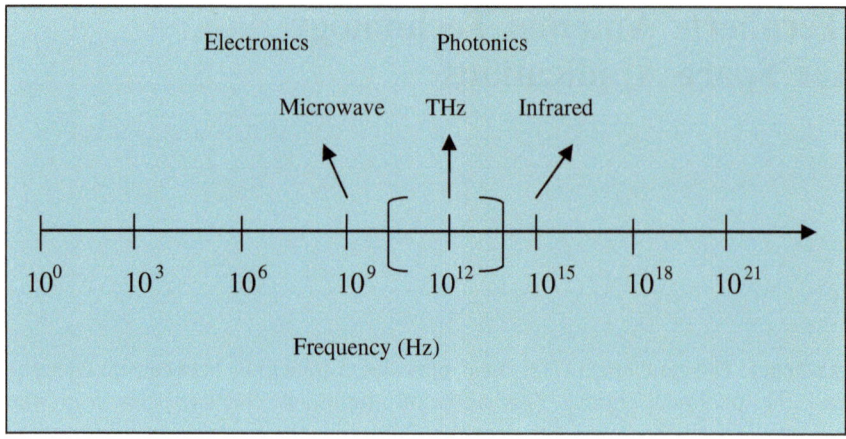

Fig. 1 The electromagnetic spectrum around the THz region

implemented around the second World War, the terahertz technology is still in its infancy and it has not reached levels of maturity to drive the market. Thus, there is a lot of research and development that has been undertaken in the last two to three decades to understand and execute this technology. The two major challenges where technology needs to be pressed upon are the implementation of compact, consistent, and cheap terahertz sources, and high sensitivity terahertz detectors. The additional complexities involved in scaling from the microwave radiation domain also need to be specifically mapped and managed (Hanson and de Maagt 2007).

This brief account attempts to give an overview of this rapidly developing technology and focuses on one important aspect: the antenna technology.

Classification of terahertz antennas is an important aspect in analyzing the trends and the future research and development avenues. The terahertz antennas can be classified according to the antenna technology undertaken to implement it, and on the basis of applications.

1.1 Characteristics and Advantages

The location of the terahertz spectrum between the microwave and the infrared regimes implies that it provides a transition between the electronic and the photonic region, thus terahertz radiation can be generated by both avenues.

Advancements in micromachining techniques and optoelectronic devices have facilitated us to manufacture the sources and receivers at the terahertz wavelength order. It is the wavelength of the radiation that characterizes the antenna size and propagation. Table 1 quantifies the terahertz domain in different units. The corresponding energy of the terahertz radiation is of the order of a few milli-electron

Table 1 The terahertz domain in different units

Frequency (THz)	Wavelength (μm)	Energy (meV)	Temperature[a](K)
0.3	1000	1.24	14.4
3	100	12.41	144.0
10	30	41.30	479.3
30	10	124.10	1440.2

[a]Energy temperature relationship : $E = k_B T$; (k_B is the Boltzmann constant)

volts. This energy does not affect biological systems and is assumed to be non-ionizing in nature. The temperatures of radiation suggest that the devices, which are semiconductor based, need to be cryogenically cooled.

The terahertz band meets the ever increasing demand for bandwidth in ultrafast wireless communication systems. Since terahertz frequency band is quite unused, researchers are more interested in the potential use of this frequency range.

Terahertz photons can be generated by changing the thermal, bending, or rotational state of a molecule. This thermal radiation, as a source of terahertz radiation interferes with the measurements. This can be used to our advantage in passive imaging to detect hot objects.

1.2 Terahertz Sources

Until the mid-1970s thermal radiation was the only source of terahertz radiation; it was used in the far infrared spectroscopic systems as a source.

In 1975, D.H. Auston at Bell Labs irradiated a biased semiconductor with a femtosecond laser pulse to generate a picosecond time varying current. The corresponding radiation contained components of the terahertz frequency band. This set up was called the 'Auston switch' and it formed the basis for the growth and development of research in this field (Fitch and Osiander 2004).

Using femtosecond lasers, an alternate method of generating terahertz pulses involves nonlinear optics. Optical rectification (Bass et al. 1962), a nonlinear optical method uses a nonlinear optical crystal which from the wide spectrum ultrafast laser generates difference frequencies in the range of 0.5–20 THz.

Photomixing involves two continuous laser pulses directed at a semiconductor material, which generates time varying current at the difference frequency inside the semiconductor, which results in terahertz radiation. This process is heavily dependent on the carrier lifetime in the semiconductor.

Radiation with a frequency of a few tenths of terahertz can be achieved by multiple harmonic generations of the microwave sources. The drawback is that a fraction of the power is lost at each stage of the process and thus the resulting output efficiency is low (One of the sources of loss is the metal waveguides at high frequencies).

Micromachining techniques have helped in adopting the vacuum tube technologies from the microwave domain. Backward wave oscillator (BWO) wherein the radiation flow is in the opposite direction to the electron beam, is the most notable tube technologies used as a source of terahertz radiation. This technology is inhibited because of its requirements of a large enough external magnetic field, a stable high voltage supply, and a terahertz modulator.

Recent developments in terahertz sources include a quantum cascade laser (QCL). In a QCL multiple quantum wells result in electrons cascading, and the transition from one well to the other results in the emission of a terahertz photon. The advantages of QCLs include significant reduction in size and considerable power output. As this is a relatively new technology, there are many challenges, such as cooling (liquid nitrogen level temperatures are required), device lifetime is low and the high costs (Tonouchi 2007; Fitch and Osiander 2004) reported a 4.4 THz, 4 mW device operating at 48 K. As of 2014, use of advanced HEMT devices (In P based), RTDs, and Schottky barrier diodes for various aspects of generation, detection, photo-mixing, etc., has been reported.

2 Applications

The driving motivation for research in this domain has been the study of our atmosphere and astronomy. But now, it has found applications in numerous fields ranging from communications and biomedical imaging and detection, etc. (Hanson and de Maagt 2007). The non-ionizing nature of terahertz radiation can be applied to various security and biomedical applications such as detection of weapons, cancer imaging, etc. Table 2 shows the trends of these applications with respect to

Table 2 Terahertz applications timeline

	2007	2010			2015
Communications	10 Gbps	40 Gbps			100 Gbps
Medical		Cancer Imaging	On site diagnosis		Fibre Coupled
Food	Inspection System			Food quality control	
Security	Imaging	Explosives	Airports, etc.		Mobile

their time frame (Tonouchi 2007). This diversification has forced the researchers to come up with feasible and novel devices for the systems, i.e., detectors, antennas, probes, etc. (Grade 2007).

2.1 Communications

With an expected total volume of mobile traffic 327 Pet bytes by 2015, the need for feasible ultrafast wireless networks is imperative. The THz band region is yet to be allocated for specific applications, frequencies greater than 275 GHz have not been allocated by the ITU and thus it can be exploited for high speed communications. Terahertz communication technology is still in its infancy, but the future looks promising (Song and Nagatsuma 2011). Figure 2 shows a secure military wireless terahertz communication network. Figure 3 shows possible wireless personal area network architecture with terahertz links as provided by the allocated spectrum chart (by the U.S. Federal communications Commission).

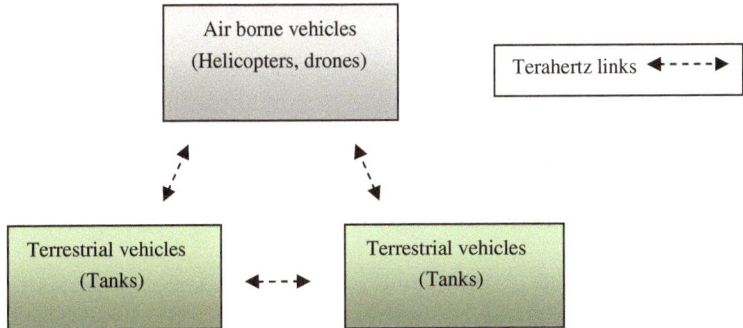

Fig. 2 Secure military communication applications

Fig. 3 Wireless personal area network

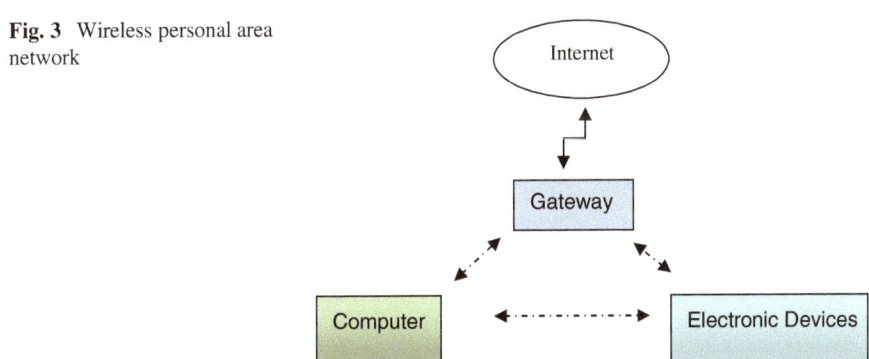

Higher bit rates can be achieved essentially by increasing the carrier frequencies; 10–100 Gbps corresponds to about 100–500 GHz (Huang and Wang 2011).

Another major advantage of frequencies greater than 300 GHz is the antenna size which reduces to about submillimeters. The implementation of these systems is now possible due to the advancements in the realization of the photonic and semiconductor devices with an operating frequency in the terahertz band. Si-based VLSI techniques for communications control (Nagatsuma et al. 2010), micro-electronics mechanical systems (MEMS) based devices and Metamaterials for antennas hold the key in feasible implementation and adoption of terahertz-based communications.

Nagatsuma et al. (2010) reported a high frequency photodiode used to generate frequencies of 300–400 GHz, with the frequency-output power relation derived by heterodyning at 1.55 μm with characteristics:

- 3 dB bandwidth is 140 GHz (from 270 to 410 GHz)
- Output power peak value was 110 μW at 380 GHz
- Photocurrent of 20 mA corresponds to 440 μW of peak output power
- Error free 2 Gbps transmission at 300 GHz (power transmitted 10 μW).

Atmospheric attenuation is highly significant for THz communications, thus its use in outdoor data transmission is not viable, whereas for indoor short range applications it is not an issue and provides an attractive option (Song and Nagatsuma 2011). Using Friss' formula given in Eqs. (1) and (2) (Schneider et al. 2012), the characteristics of short range links can be approximated.

$$P_r = P_t + G_t + G_r + 20 \cdot \log\left(\frac{\lambda_c}{4 \cdot \pi \cdot d}\right) - L_{\text{additonal}} - \{\alpha \cdot f_c \cdot d\} \qquad (1)$$

$$\text{SNR}(dB) = P_r - \{N_0 + 10 \cdot \log B + \text{NF} + M\} \qquad (2)$$

P_r is the power received, P_t is the power at the transmitter end, G_r and G_t correspond to the antenna gains of the receiver and the transmitter; α is the attenuation due to the atmosphere, d is the distance, and λ is the wavelength; NF is the noise figure of the systems, M is the noise margin, and B is the bandwidth of the system.

2.1.1 5G Cellular Networks

Terahertz band can be used for high speed data transmission within a range of 10 m. This coverage area consists of small cells of cellular networks. Terahertz communication is applicable in the indoor as well as outdoor environments with stationary and mobile users. Terabit wireless local area networks (T-WLAN) can provide flawless communication between high speed fiber optical links, and personal laptops and tablets. Wired and wireless links enjoy the same speed in terahertz communication. This will enable the users to experience the bandwidth-intensive applications such as wireless data distribution in data centers, high quality video conferencing, etc., in the indoor scenarios. Another peculiar feature of the terahertz communication is the secure communication links for military and defense applications. Since the THz Band frequencies experience high atmospheric attenuation and signal losses, large antenna arrays are required to overcome the coverage area limits. The resulting narrow antenna beams reduced the eavesdropping probability. Further, the signal jamming attacks can be overcome by various spread spectrum techniques.

2.1.2 Challenges in Communication Systems

There is a need to develop novel solutions at the different levels of the OSI model for the communication networks, and to adopt new communication paradigms involved in a THz system (Akyildiz et al. 2014). Other challenges for the future THz communications include interference with astronomical telescopes and the other technical problems of beam scanning and packaging.

Due to very high sensitivity of radio telescopes any sort of interaction of future THz communication systems needs to be minimized. This particular problem has been studied by the ITU-R Study Group 7. The assumptions considered for the worst cases are given as:

- The empirical equation for output power from the interfering transmitter is given as $P_T = 0.01 * (1000 - f_{GHz})$ dBm
- Interfering transmitter is directed toward the radio telescope
- The interfering transmitting antenna is of 30 cm diameter which corresponds to 58 dBi at center frequency of 275 GHz
- Radio telescope is directed toward the sky
- The antenna gain of telescope in the direction of interfering transmitter should be 0-dBi
- The interfering transmitter and the radio telescope are at 3000 m altitude
- Radio astronomy protection criteria: As defined in ITU-R Recommendation RA.769

Following these assumptions, the minimum distances for the interfering signal to be negated at the radio telescope have been given in the following Table 3 (for frequencies between 275 GHz and 3 THz).

Table 3 Minimum distances for the interfering signal to be negated at the radio telescope

Frequency range	Minimum distances (kms)
275–320 GHz	55
335–360 GHz	26
380–445 GHz	14
>450 GHz	6

2.2 Detection and Imaging

Electro optic effect wherein the electromagnetic wave passing through a nonlinear crystal polarizes in proportion to the applied electric field is the fundamental principle in detecting terahertz pulses using nonlinear optical crystals. The terahertz pulse and the laser pulse are passed through the crystal at the same time so that the polarization of the latter changes in proportional with respect to the terahertz electric field. The time delay between the terahertz pulse and the femtosecond laser pulse gives the time dependence of the electric field of the terahertz pulse. A Grischkowsky antenna, also known as a photoconductive antenna, discussed in detail later, is often used as a detection component in time domain spectroscopy (Fitch and Osiander 2004).

Imaging applications using terahertz radiation are central to the research undertaken in this domain. Important characteristics of terahertz imaging are its ability to provide imaging through dielectrics, with an optimum level of resolution. Resolution is a function of the beam diameter at the corresponding terahertz wavelength. The inferences from the important amplitude and phase information can be used to find mechanical defects, and in medical applications, security applications, etc.

2.2.1 Terahertz-TDS System

The fundamentals of time domain spectroscopy involve the measurement of the electric field as a function of time; applying Fourier transform to this function extracts the frequency spectrum, which gives the corresponding amplitude and phase information. Few of the important characteristics of TDS are:

- High signal-to-noise ratio (SNR)
- Coherent detection ensures that measurements are not affected by background thermal radiation.

  The working principle of TDS:

- A femtosecond laser using nonlinear optics generates terahertz pulses. This is followed by splitting the gating beam. A fraction of this beam is then combined in an electro-optical crystal with the terahertz pulse.
- Measurement of the polarization shift is undertaken using polarization optics and 2 photodiodes (Electro-optical effect).
- The electric field as a function of time and hence the relevant amplitude and phase information can be extracted by tuning the time delay between the terahertz pulse and the laser pulse.

  Figure 4 shows the block diagram representation of a TDS system. The TDS can be used in many detection applications such as detection of chemical species, explosives, bio-agents, absorption spectroscopy, and impulse ranging to name a

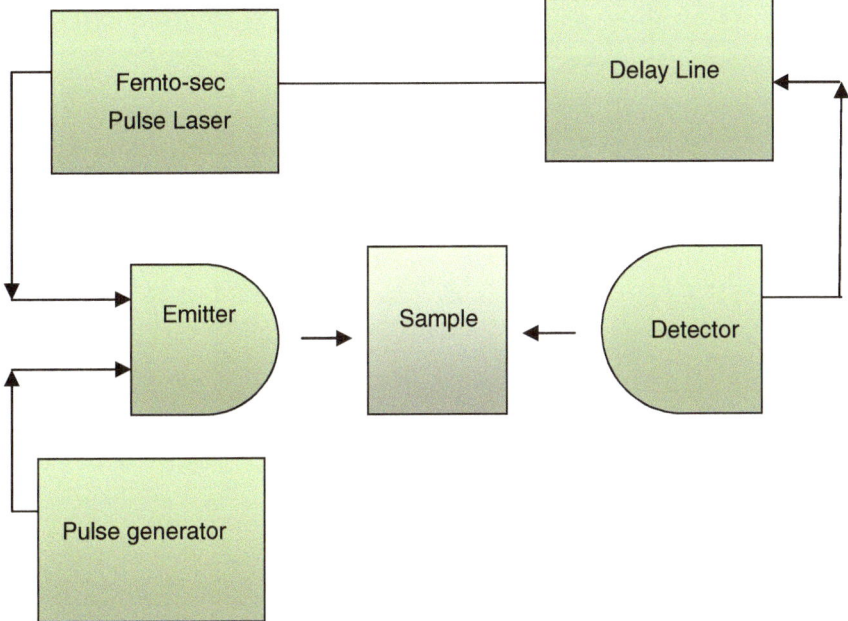

Fig. 4 Block diagram of terahertz-TDS spectroscopy system

few. One of the methods of imaging is using a photoconducting emitter which employs the time domain set up. As the object is scanned, each pixel of the image has its corresponding time domain waveform and frequency spectrum (Fitch and Osiander 2004).

2.2.2 Frequency Domain Spectroscopy System

Also known as continuous wave spectroscopy, it is an alternative to the time domain set up. These systems can be tuned continuously over the terahertz range. Photomixing techniques combine two single color laser diodes which illuminate a narrow gap between biased semiconductor materials which produces difference frequency terahertz radiation due to the photocurrent. This is then split into emitter and detector components in the coupler. The continuous wave that is received at the photoconductor is mixed with the coupler optical signal to attain coherent detection. Figure 5 shows the block diagram representation of a frequency domain spectroscopy system.

This set up is more feasible as no femtosecond laser is required and the tuning flexibility gives an improved resolution (www.cryotronics.com) (Tonouchi 2007). Usually for detection purposes the time domain set up is used, whereas the frequency domain set up is used for imaging purposes.

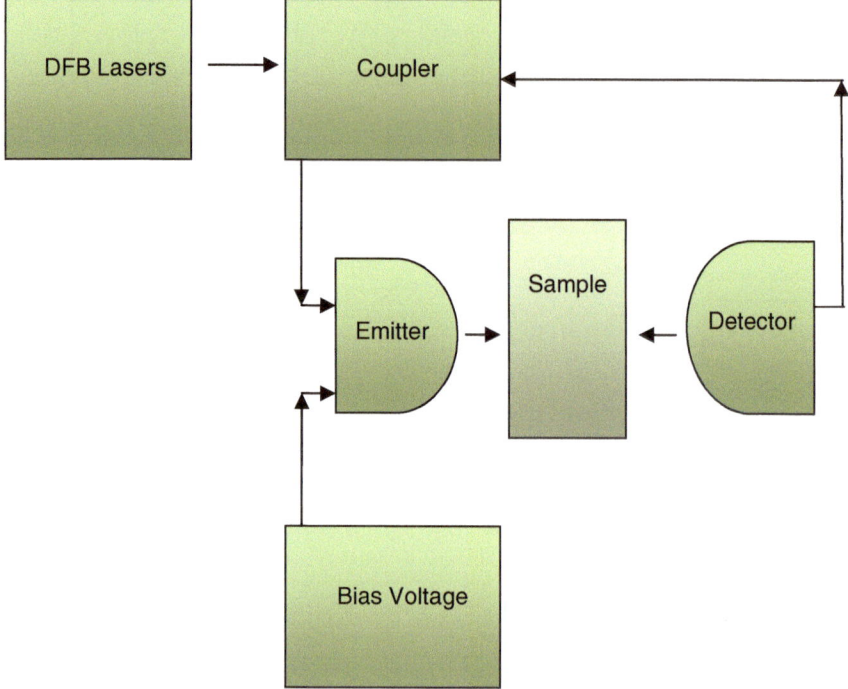

Fig. 5 Block diagram of terahertz frequency spectroscopy system

3   Terahertz Antennas

Whether for imaging, communications, or astronomical purposes, all the various implementations of an antenna are a function of its size, bandwidth, feed elements, and the aperture (Siegel et al. 2006). Scaling of the existing technologies to adapt to the terahertz region leads to inefficient designs and does not give optimum solutions. Table 4 compares the characteristics of microwave antennas and the terahertz antennas.

3.1   *Photoconductive Antennas*

Since conductors with their small skin depths have greater conductive loss, photoconductive antennas are an alternate viable option (Li and Huang 2006). Energy inputs such as pulse lasers with short periods to the semiconductor generates excitations; on biasing, these excitations form a current. Varying the energy input varies the intensity of the excitations, thus forming a time varying electric field

Table 4 Important differences between millimeter and submillimeter antenna technology

Constraints	Microwave antennas	Terahertz antennas (photoconductive)
Source	Transmission line	Laser
Substrate	Dielectric media <<wavelength	Substrate thickness is comparable to the wavelength
Impedance matching	Easier to achieve	Relatively difficult to attain
Biasing	Not required	Required
Fabrication	Easier	Harder
Computer aided design	Available	Specific to THz regime not available

which in turn results in radiation. The geometry is an important factor for analysis of terahertz antennas as the substrate thickness cannot be ignored in this region, since it is comparable to the wavelength. Greater electric fields formed by an indented antenna as compared to a conventional dipole antenna provide better performance characteristics. Figures 6 and 7 shows the difference in the circuit modeling of a conventional antenna and a photoconductive antenna (Fig. 8) respectively.

Li et al. (2010) demonstrated a photoconductive horn antenna (Fig. 9) with the aim of unidirectional radiation as compared to the traditional dipole antenna which radiates in all directions. The factors that need to be considered are the size of the

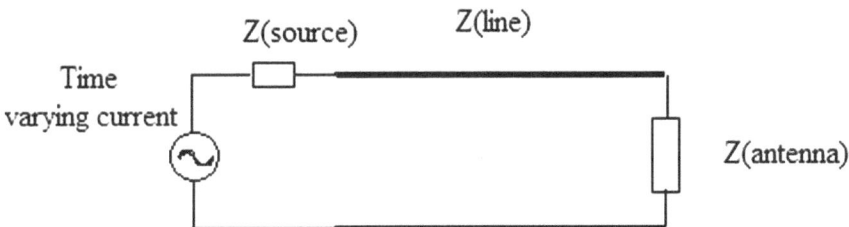

Fig. 6 Circuit diagram for a conventional antenna

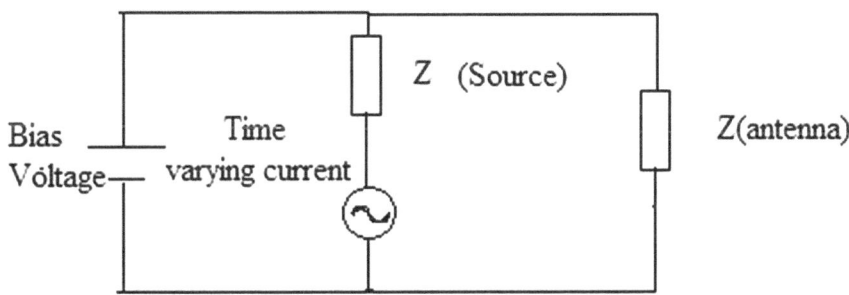

Fig. 7 Circuit diagram for a photoconductive antenna

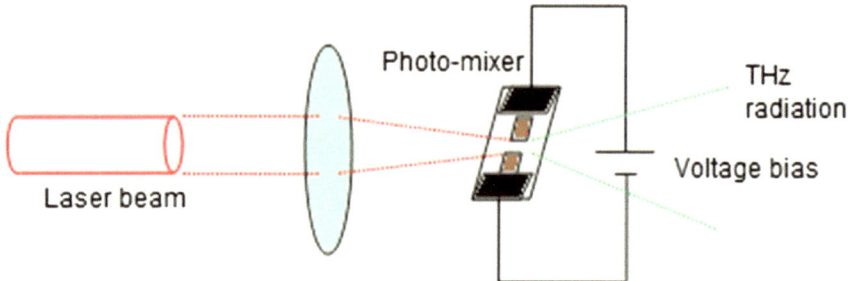

Fig. 8 Schematic of a photoconductive antenna

photoconductive emitter gap, indentations as compared to traditional dipole electrodes, substrate thickness, and the back wall membrane to ensure unidirectional
radiation. Li et al. (2010) also use indium tin oxide (ITO) for the back wall of the
horn antenna. Biasing the antenna is another problem that can be addressed by
extending the substrate out of the horn, and then adding strip lines and metal pads to
connect the electrodes to the voltage source. The model can be simulated in
FEM-based software and can be optimized thereon.

Huang et al. (2011) analyzed a photoconductive antenna and derived a new
efficiency expression as a function of its various parameters. Its low efficiency
continues to be a drawback as it results in low power terahertz waves. It is noted
that the overall efficiency is the product of the efficiencies of the three components,
photoconductive element, its impedance, and the radiation as seen in Eq. (3)
(Huang et al. 2011). The laser to electrical efficiency manifests itself as it is not only
the most dominant of the three, but also the worst. For a time domain system, the
photoconductive material is the most significant element, but for the continuous
wave systems the impedance matching is the most dominant. The topic of photoconductive terahertz antennas is still in its infancy at the moment, eventhough the
efficiencies are low, the topic is being researched thoroughly as it is an attractive
option for the practical and feasible implementation of terahertz sources.

$$\eta_{le} = \frac{P_e}{P_l} = \frac{e \cdot V_{\text{bias}}^2 \cdot \mu_e \tau^2 \cdot \eta_L \cdot f_R}{h \cdot f_L \cdot l^2} \tag{3}$$

η_{le}	Laser to electrical efficiency
P_l	Power output of laser
P_e	Electrical Power output
V_{bias}	Voltage Bias
μ_e, τ	Property of photoconductive material
η_L	Efficiency of illumination
f_R	Laser repetition frequency
f_L	Laser frequency
l	Gap length

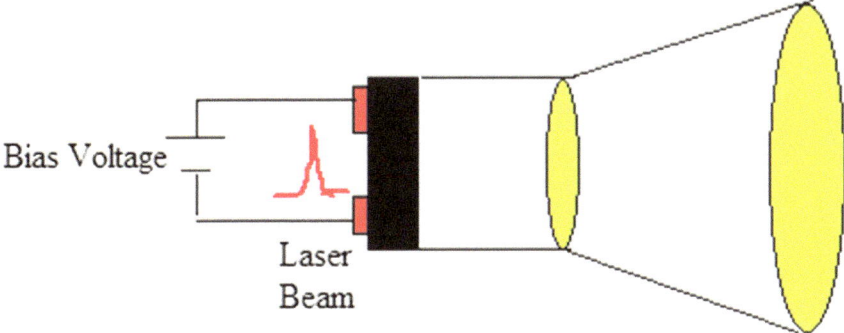

Fig. 9 A schematic of a photoconductive horn antenna

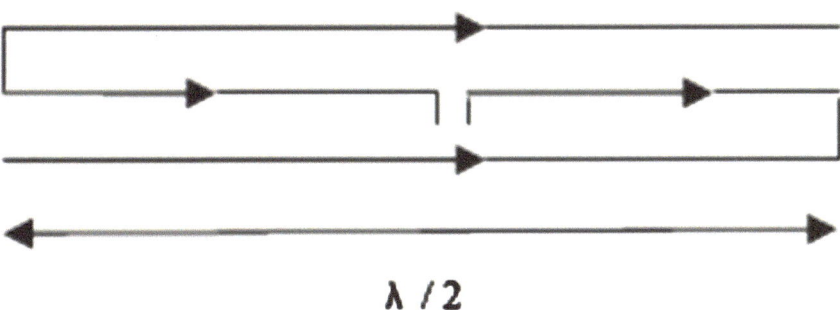

Fig. 10 Three wire folded dipole antenna

The lack of feasible terahertz sources has limited the hindered direct applications of terahertz technology in various areas (Moon et al. 2005). One of the problems related to a photoconductive terahertz source is the impedance mismatch between the photomixer and the antenna which can be canceled by increasing the resistance offered by the antenna. Using an equivalent circuit model for the photomixer loaded with the antenna impedance, Moon et al. (2005) showed that half wavelength antennas with greater resistances give a larger power output as compared to a full wavelength antenna. A 3 wire folded dipole antenna is shown in Fig. 10.

The radiation power peaks, proportional to the resistance offered by the antenna, are observed to occur at a frequency slightly lower than the resonant frequency because of the unaccounted capacitances between the wires; this can be negated by tuning the dimensions of the dipole accordingly.

3.2 MEMS-Based Antennas

There is a growing interest in the use of MEMS-based devices and new materials such as carbon fiber reinforced polymers (CFRP) which give terahertz antenna design a new dimension in order to achieve the desired characteristics. Integration of the functional elements provides better transmission, radiation with a higher order of reliability and compactness. These are among the few challenges faced by researchers in the implementation of various terahertz components.

Yong et al. (2006) presented a pyramidal horn antenna, pyramidal single ridged, and a double ridged antenna using MEMS technology with an operating frequency range of 0.3–1.5 THz. They approximate the frequency range of a single ridged waveguide to be around 0.35–1.48 THz. The use of MEMS technology for designing and manufacturing the radiation components, and the feed elements of the antenna for gaining popularity. Not only does MEMS technology offer a higher degree of integration, but it also results in a greater efficiency and enhanced reliability. The author also lists the merit of single crystalline silicon (SCS) as the base material structure. Characteristics such as tuning of material properties by addition of dopants, nanometer level surface chemical polishing, good mechanical properties (elasticity, rigidity, etc.), and anisotropic etching makes SIS a very good basic material structure. Yong et al. (2006) also recommend a ridged waveguide as compared to a rectangular waveguide to expand the frequency band. Simulation of the array can be carried out using FEM-based software, fabricate the structure using the bulk micromachining technique on the Si wafer. The antennas are designed by exploiting the anisotropic nature of silicon. The waveguide and the horn are micromachined separately, and are then joined to form the antenna. The elements of the array are spaced accordingly to attain low side lobe levels. The uses of pyramidal horn arrays include high directivity applications, and that of the ridged horn arrays are for imaging a broad frequency range.

Bowen et al. (2006) demonstrated a monolithically fabricated novel horn antenna with a frequency of operation of 1.6 THz. The recent advancements in MEMS-based techniques have allowed us to manufacture complex integrated components. Limitations in fabrication also limit the characteristics of the antenna; the maximum height of the structure is a function of the thickness of the photoresist in a single layer.

3.3 Photonic Band Gap-Based Antennas

Metallic conductors due to their non-ideal nature (dielectric absorption, larger skin depth) result in high losses at terahertz frequencies. This degraded performance can be tackled by introducing novel waveguide structures which reduce losses at higher frequencies (Sánchez-Escuderos et al. 2011).

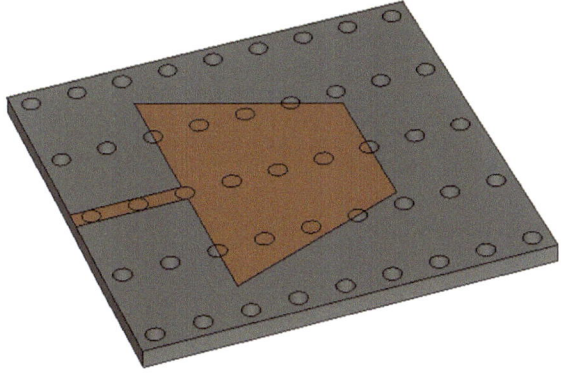

Fig. 11 Schematic of a PBG antenna

Controlling the propagation of electromagnetic waves through a medium, allowing only the desired frequency bands to pass; we can control the transmitted output. This band pass filtration can be realized using photonic crystals. Materials with dielectric constant patterned with a particular periodicity to restrict the passage of certain frequencies, thereby forming a photonic band gap (PBG) are called photonic crystals as shown in Fig. 11. High free space path loss decreases the efficiency of transmission, thus waveguide elements are crucial fundamental components for terahertz transmission. The efficiency of a rectangular waveguide is a function of roughness and conductivity of its side walls. The transmission efficiency in case of a PBG waveguide is improved as the sides are an array of columns not closed by metal (Liu et al. 2010).

Wideband, multifunctional devices are the need of the hour. Integration of receiver and emitter components would ensure the optimum usage of this technology. The difficulty lies in implementing the antenna as an easy-to-integrate component on a semiconductor substrate while maintaining optimum radiation efficiency. High dielectric constant of the substrates results in a significant loss in the desired power output. This substrate loss can be negated by using a PBG structure (Fig. 12). A PBG waveguide consists of a dielectric substrate sandwiched between two metal plates; the substrate is perforated with air holes. PBG technology provides efficient control and the capability to guide electromagnetic waves, this important characteristic has given us new avenue in planar antennas as compared to the conventional types.

Sánchez-Escuderos et al. (2011) presented the design of electromagnetic band gap waveguides which result in low attenuation in the terahertz region of operation. In a multilayer rectangular waveguide structure, the conductor encloses the two layers of dielectrics. The material with greater permittivity lines the side walls. To achieve wideband transmission, the dielectrics may be chosen such that the difference in the layer permittivity is large. The multilayer waveguide structure does not eliminate the losses in the upper and lower walls, thus three-dimensional EBG

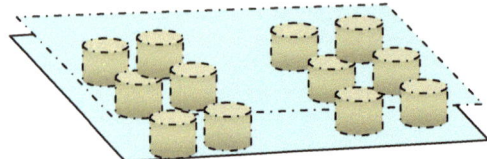
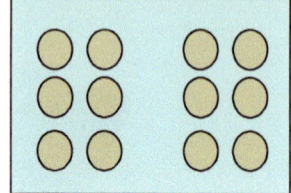

Fig. 12 Schematic of a PBG structure

Table 5 Type of waveguide and corresponding attenuations

Type of waveguide	Attenuation (dB/cm)
Rectangular	0.9
Multilayer rectangular	0.7
Woodpile structure	0.4

structures can be used in order to eliminate the losses in all directions. Yablonovitch first put forth a proposal for a practical three-dimensional EBG in 1993. De Maagt et al. (2003) examined the PBG technology for microwave and terahertz antennas using a woodpile (layer by layer) PBG structure. The woodpile structure forms the basis of the three-dimensional EBG waveguide.

The dielectric periodicity is maintained in all three dimensions, thus restricting the propagation of the waves. Placing the dielectric rods in the air medium periodically in all directions forms a woodpile structure. The constraints in all directions ensure that the attenuation is significantly reduced. Table 5 shows the simulation results of the design proposed by Sánchez-Escuderos et al. (2011) (at 0.950 THz).

A horn antenna can be further designed by maintaining the woodpile structure and increasing the width of the multilayer structure. The resultant field is confined in the central region, thereby increasing the gain and reducing the attenuation (Sánchez-Escuderos et al. 2011).

Xu et al. (2013) exhibited a high directivity radiation pattern for a compact Cassergrain reflector antenna with radiating and receiving frequency of 0.33 THz. The estimate of the antenna characteristics can be done using folded optics in the design of a feeding horn. In their analysis they report a far field directivity of 50 dB and a beamwidth of 0.5°. Solving Maxwell's equations for large electrical apertures is cumbersome and using path theory based on the geometry provides a good approximation to calculate the antenna performance. Near field distribution is extrapolated to get the far field pattern and the RMS surface error can also be taken into consideration to evaluate the corresponding directivity. This design can especially be employed for detection and imaging purposes, because of its high directivity.

The high speed wireless communication in space aviation is made easier with the design of THz devices. Since atmosphere is absent in the space, the issue of atmospheric attenuation with THz waves would not affect the space

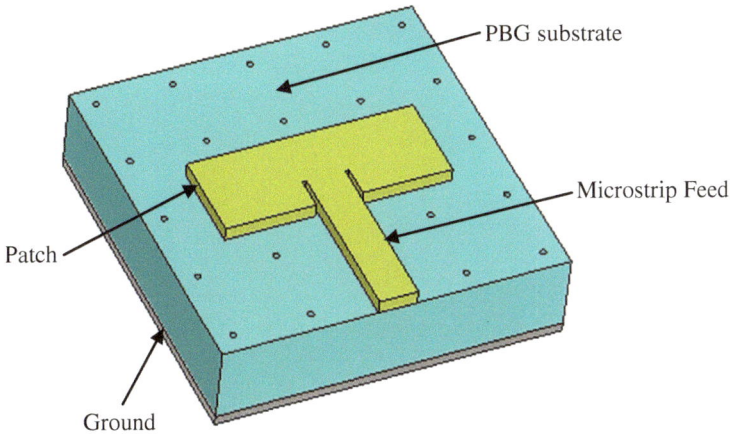

Fig. 13 Schematic of a high gain PBG-based microstrip antenna

communication. High gain antennas which can mitigate the free space loss need to be implemented. A microstrip patch antenna designed on a PBG substrate exhibit better return loss, high gain, and high directivity. The advantages of using microstrip antennas are light weight, low cost, and reduced fabrication complexity. These patch antennas can also be easily integrated onto conformal surfaces. The schematic of a high gain PBG-based microstrip antenna is as shown in Fig. 13. The return loss (38. 54 dB) and radiation characteristics (gain: 13.5 dB) of the antenna are as shown in Figs. 14 and 15 respectively (Danana et al. 2014).

3.4 Arrays

Since the space loss for terahertz radiation is significant, the use of high gain and high directivity antennas is imperative. This can be implemented through a phased array. Individual components of the array are phase matched thus increasing the gain and the directivity. Integration of individual components to form an array is another distinct advantage of micromachining techniques. When it comes to space applications, mass and volume of the instruments are also essential design considerations. Integration of antenna arrays coupled with detectors are ideal for such applications (Goldin et al. 2002). In order to maintain optimum levels of sensitivity and resolution keeping in mind the importance of compactness, low mass focal plane antenna arrays provide an elegant alternative (Chattopadhyay 2010).

Array systems are limited by the aperture number. Higher directivity can further be achieved using a leaky wave. A dielectric super layer between the feed and the iris allows the waveguide the wave modes which otherwise would not be available (Llombart et al. 2008; Chattopadhyay 2010).

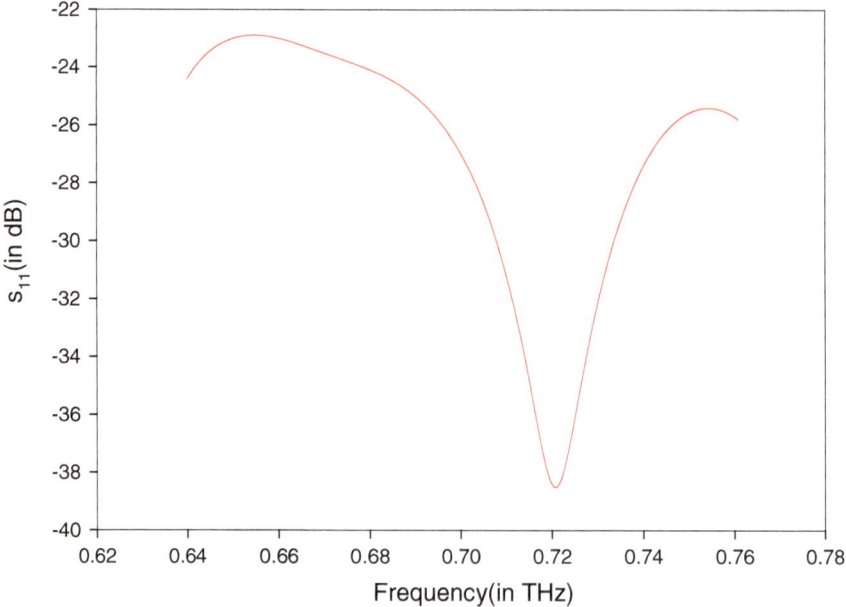

Fig. 14 Return loss characteristics of the PBG-based microstrip antenna

Llombart and Chattopadhyay (2010) further adopted this concept by attaching a microlens to enhance the directivity. The restriction of the field to a narrow angle implies that the entire curvature is not required, thereby easing the fabrication on the bulk silicon substrate.

The geometry of such integrated microlens antenna arrays includes a waveguide which is further integrated with Schottky mixers, etc., which act as feed elements to the lens through a leaky wave or a PBG structure which are also matched to the waveguide (Llombart et al. 2013). These high gains, high directivity integrated array structures are efficient in imaging, heterodyne, and planetary applications.

3.5 Nanostructures

Considerable advancements in nanotechnology have taken place since Richard Feynman, in his landmark speech, "There is plenty of room at the bottom" initiated research in this foray. The advent of nanostructures such as graphene, carbon nanotubes, and metamaterials coupled with MEMS concepts have given antenna design for the terahertz regime a different avenue. A consequence of miniaturization to the nanoscale is that the underlying physics is now governed according to quantum mechanical considerations. Nanostructures such as carbon nanotubes (CNTs) have been analyzed as possible dipole nanoantennas and nanoribbons as

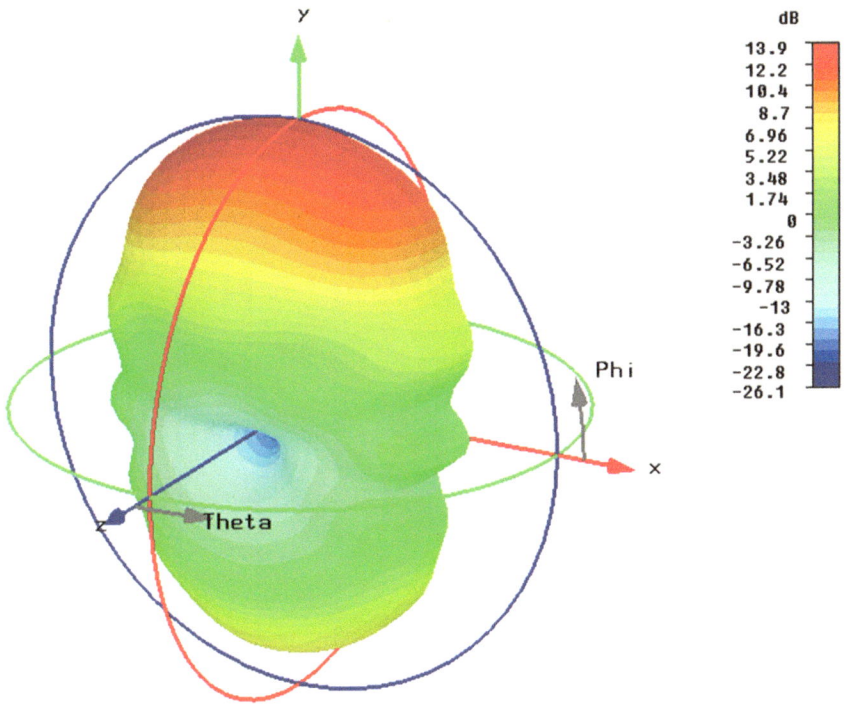

Fig. 15 Radiation characteristics of the PBG-based microstrip antenna

patch antennas. The waveguide and transmission line properties of CNTs are a function of the geometrical properties of the nanotubes. Jornet and Akyildiz (2013) presented a model for the impedances (Quantum Resistances, Capacitances, and the Kinetic Inductance) of CNTs. The technology has not matured yet and energy consumption, geometrical complexities are few of the hurdles researchers face. Llatser et al. (2012) reported similarities in the radiation measurements between a graphene-based nanopatch antenna and its metallic equivalent, thus proving that these nanostructures can indeed be used in antenna design. Figure 16 shows a graphene-based nanopatch antenna array used as a reflector to achieve the desired directivity and gain (Carrasco et al. 2013).

3.6   Graphene-Based Antennas

The antennas can be patterned on graphene for terahertz frequencies to achieve good radiation characteristics and high gain. In graphene-based terahertz antenna arrays (Fig. 17), radiation pattern of an antenna array can be varied by controlling the gate voltages acting on each antenna. This eliminates the need for switches or

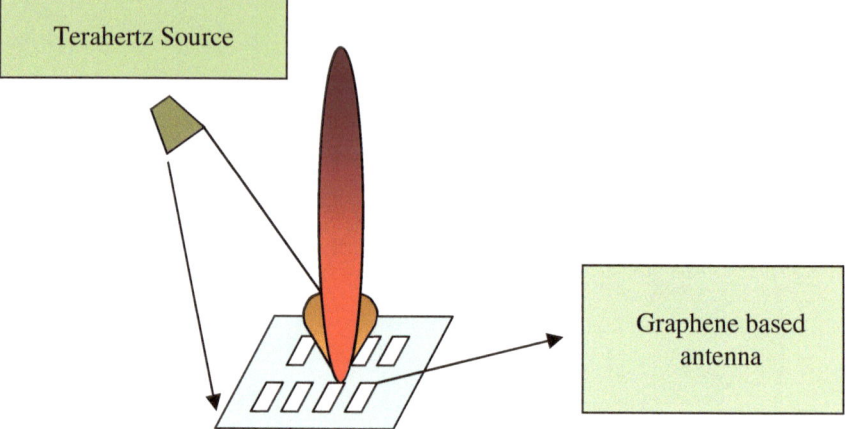

Fig. 16 Graphene-based reflector array

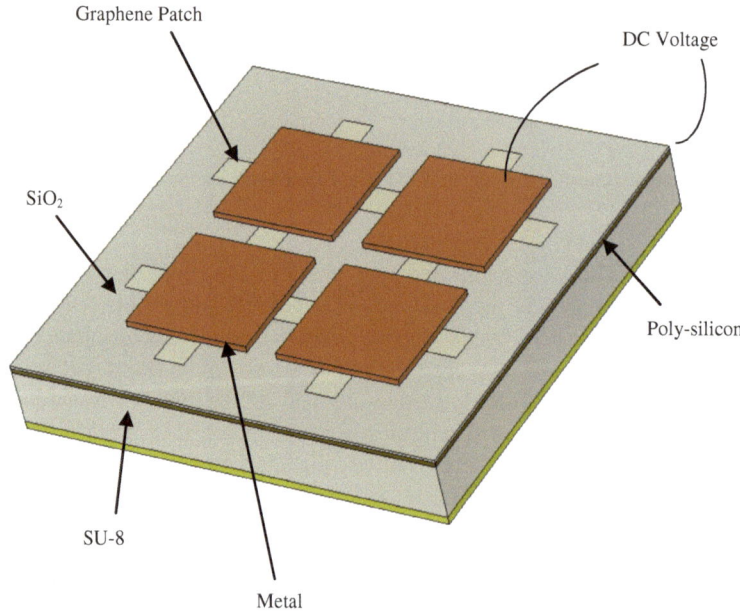

Fig. 17 Graphene-based reflector array

phase shifters for beam steering. The value of absorption is very low in graphene-based terahertz antennas. These antennas also show good gain and radiation efficiency in terahertz frequency band. Since the size of the antenna is very small for terahertz frequencies, it will occupy only a small area in the graphene layer. The desired radiation where the THz electromagnetic energy concentrated in main lobe and with reduced sidelobes could be achieved easily by modifying the gate voltages. (Dragoman et al. 2010).

3.7 Leaky Wave Antennas Using Graphene-Based High Impedance Surfaces

A planar array with periodic unit cells can be considered as a high impedance surface (HIS) which can manipulate the propagation of electromagnetic wave. Graphene can be adopted as a radiation surface which can support THz waves with improved radiation performance.

These graphene-based HIS can be implemented into the design of reconfigurable leaky wave antennas (LWA). Graphene metasurfaces can be easily tuned by varying the applied electric field bias. Reconfigurability can be achieved for resonance as well as for beam direction (Chen et al. 2015).

3.8 Substrate Integrated Antennas

The substrate integration technology converts nonplanar antenna structures into their planar forms. Advanced microfabrication techniques are adopted for the design of terahertz antennas. Some of the substrate integrated antenna structures used in THz technology are slot array, reflector, horn and leaky wave antennas. The substrate integrated reflector antennas exhibit high gain and wide bandwidth. But large size, fabrication methods and cost of fabrication limits the use of such reflectors. Substrate integrated patch antennas overcome all the limitations of reflectors but gain and bandwidth are very small. Patch antennas are easier to fabricate, light weight and of low cost compared to any other substrate integrated antenna structures. Though substrate integrated horn antennas have the advantage of wide bandwidth, the fabrication methods are too difficult compared to former substrate integrated antennas. Dielectric lens antennas exhibit better bandwidth than patch antennas. But their gain is very less similar to patch antennas. The ease of fabrication, cost, and size are the other salient features of dielectric lens antennas. (Wu et al. 2012).

3.9 THz Antennas with Artificial Dielectric Superstrate

The radiation performance and efficiency of THz antennas can be enhanced using artificial dielectric layers (ADL) over the antenna structure. ADLs are periodic metallic patches which are of non-resonant nature. The high dielectric losses and limited bandwidth of on chip integrated antennas can be mitigated with AD superstrates. The non-resonant periodic nature of the ADL leads to improved bandwidth in THz antennas. The effective electrical permittivity and front to back ratio will be increased with the addition of ADL as superstrate in THz antennas. Aluminum can be used as the metal, and silicon oxide as the dielectric substrate in the fabrication of THz antennas. The ADL of specified thickness are carefully deposited on the antenna structure. ADL exhibit high surface wave efficiency due to the change in properties of the material depending on the direction of incidence of wave. The dielectric permittivity is high for normally incident waves and low for waves radiated parallel to the dielectric slab. (Syed et al. 2015).

4 Antenna Measurement and Testing

There is a need to assess decisive parameters such as bandwidth, directivity, etc., for various antennas subject to the required applications. Critical improvements in the design and characteristics of antennas and antenna arrays are essential in order to realize the desired requirements. Terahertz antenna characterization of bandwidth, impedance, polarization, etc., has not yet been methodically structured and it continues to be a major research challenge (Akyildiz et al. (2014).

4.1 Characterization

Khiabani et al. (2013) have demonstrated a new procedure to characterize terahertz photoconductive antennas. The detecting signal is characterized by taking into account the receiver effects using full wave electromagnetic solvers such as FEM and FDTD-based software. Emission characteristics are then observed by varying the system parameters. This method can be applied to test the antenna before it is fabricated at high costs. Conventional characterization of a planar antenna involves approximating the antenna as a Hertzian dipole, and then calculating its radiation field to a good estimate. Numerical solutions of Maxwell's equations and the carrier rate equations using finite difference time domain method does provide a more accurate solution, but this method is cumbersome. A full wave electromagnetic solver provides an alternative to these methods. There is a significant variation in the analysis of terahertz antennas as compared to the microwave antennas, because of the optoelectronic characteristics of the terahertz antennas as a result of the response of the photoconductive material to optical excitation.

Optoelectronic analysis estimates the photocurrent density and the gap conductance. The Drude-Lorentz model is adopted, with the assumption that the gap is small enough so as to maintain a uniform current distribution. The current density is found to be a function of the generated photocarrier density, and the carrier velocity as seen in Eqs. (4), (5), (6), and (7), (Khiabani et al. 2013).

The gap conductance depends on:

- The properties of the source which include the power, frequency, and the duration of the pulse.
- The photoconductive material properties such as mobility, carrier lifetime, etc.
- The geometry of the antenna.

According to the maximum power transfer theorem, the conductance of the gap is taken equal to the source conductance to ensure maximum output power. The magnetic field is the ratio of the Fast-Fourier transforms of its electric field and the current; this is the transfer function of the antenna. The Fourier transform of the current density is proportional to the Fourier transform of the radiated terahertz electric field and the photocarrier density. Simulations are done subject to these conditions.

The variation in the rise time is achieved by changing the pulse duration. Shorter laser pulses are required for broad bandwidths. The amplitude of the radiation is proportional to the carrier lifetime. There is a trade-off between the signal amplitude and the bandwidth of the antenna. Intuitively, one can guess that an increase in the gap decreases the amplitude of the signal, the reason being the decrease in the optical power density in the gap thus reduces the density of photocarriers which in turn reduces the photocurrent. The geometry of the antenna is constrained by the desired output and the device break down threshold. This method helps us in the overall analysis of the antenna performance.

$$\frac{dN(t)}{dt} = -\frac{N(t)}{\tau_c} + \frac{\alpha I(t)}{h\nu} \tag{4}$$

$$\frac{d\nu(t)}{dt} = -\frac{\nu(t)}{\tau_i} + \frac{e}{m^*}\left(\frac{V_{bias}}{L} - \frac{p_{sc}}{\eta\varepsilon}\right) \tag{5}$$

$$\frac{dp_{sc}}{dt} = -\frac{p_{sc}}{t_r} + J(t) \tag{6}$$

$$J(t) = e \cdot N(t) \cdot \nu(t) \tag{7}$$

Generated photocarrier density is denoted as $N(t)$, $I(t)$ is the laser pulse density, optical absorption is given by α, where $h\nu$ is the energy of the photon, carrier velocity is denoted as $\nu(t)$, p_{sc} is the space charge polarization; V_{bias} is the bias voltage, L is the antenna gap length, geometrical factor is denoted as η, ε is the permittivity of the material, τ_r and τ_c are the relaxation time and the recombination time respectively.

4.2 *Terahertz Antenna Testing*

Räisänen et al. (2010) reviewed the measurement techniques for terahertz antennas. Figure 18 shows a conventional far field measurement set up. The conventional method of far field measurements of electrically large antennas is not a viable option in the terahertz domain.

The submillimeter nature of terahertz radiation implies that the far field distances measure up in kilometers. Moreover, since terahertz radiation undergoes large attenuation over these distances, such measurements would not give any information. This issue can be handled using compact test ranges, wherein a collimating element reflects plane wave fronts at the measuring end. The accuracy in manufacturing imposes a major challenge in the implementation of these collimating elements.

Collimating elements of the compact antenna test ranges (CATRs) generate the conditions for far field measurements, i.e., they receive the spherical waves from the antenna and transmit plane waves at the antenna under test. Most commonly used collimating elements are a reflector, a lens, and a RF-hologram. Quite zone fields (far field conditions) typically require peak to peak amplitude of 1 dB and a phase variation of 10°, (Räisänen et al. 2007).

The measurement accuracy is a characteristic of the antenna under test and the collimating element. The accurate phase measurement makes the verification of the quite zone field challenging. Reflector CATRs (Fig. 19) require precise surface and shape accuracy. Reflectors provide a wideband of operating frequencies; the range is constrained by the surface accuracy (upper limit) and the size of the reflectors (lower limit). Lens CATRs (Fig. 20) are low permittivity dielectric lens structures which can be used for collimating beam.

The RF hologram structures (Fig. 21) due to their planarity are easier to manufacture (Räisänen 1998).

Fig. 18 Conventional far
field measurements

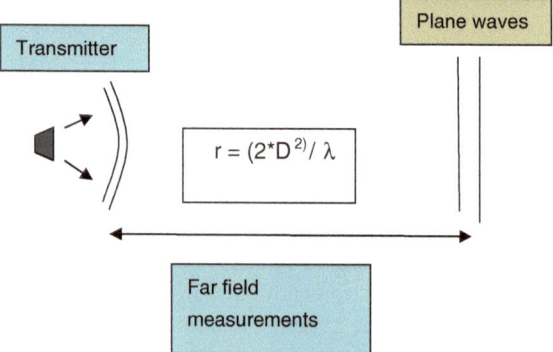

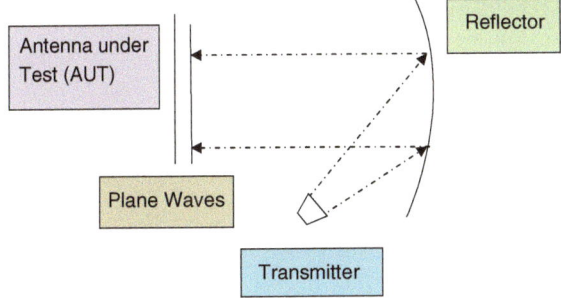

Fig. 19 Compact antenna test range for far field measurements (Reflector based)

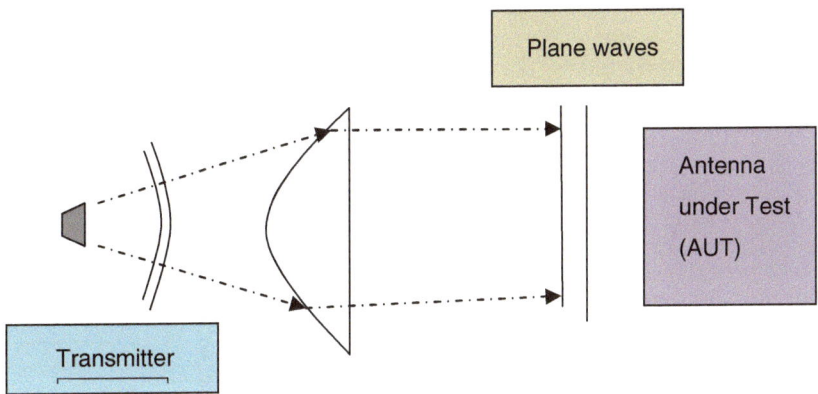

Fig. 20 Compact antenna test range for far field measurements (Lens based)

The holograms can be classified into two main categories:

- Amplitude or phase holograms
- Reflection or transmission holograms

The side view of a typical amplitude hologram is shown in Fig. 22. Near field measurements engage a probe in a particular fixed geometry which scans the near field of the AUT. This predefined geometry is usually planar, spherical, or cylindrical. The challenges associated with near field measurement techniques include accurate phase measurements, scanning the probe antenna with respect to the geometry, and the huge number of samples required to satisfy Nyquist criterion. Both CATRs and near field measurement techniques provide the advantages of indoor measurements and controlled environments.

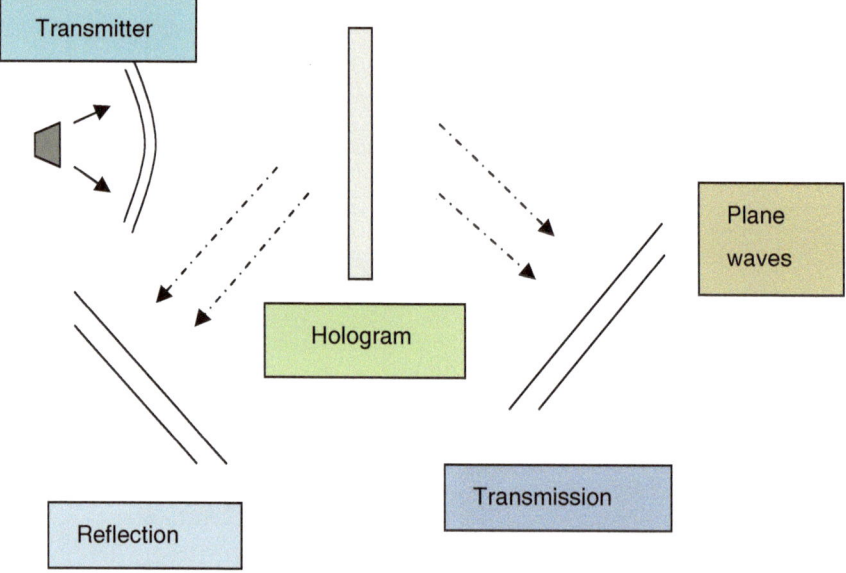

Fig. 21 Compact antenna test range for far field measurements (Hologram based)

Fig. 22 Side view of an amplitude hologram

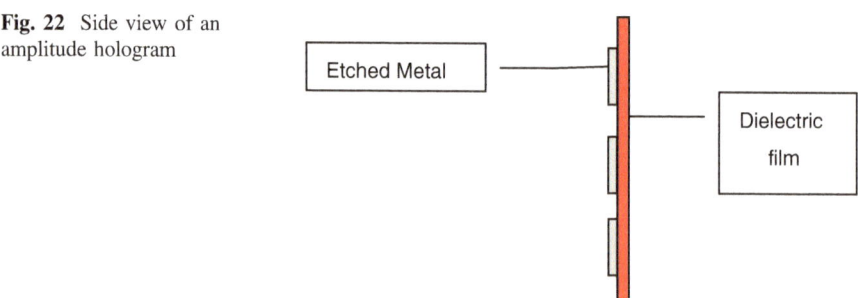

5 Space Applications

Communications, detection, and imaging are all quintessential components of a space instrument system. Heterodyne receivers detect the terahertz signals in space; antenna arrays can be implemented for imaging purposes, higher data rates for space communications can be realized through the advancements in terahertz communication systems. The research in antenna technologies discussed has culminated from the space and astronomical applications point of view.

An example of indoor communications application of the THz systems is for the interior WLANs for the Space Station modules. Additional gain is required to compensate for the atmospheric losses. Jih et al. (2013) approximate that a 5″

Table 6 Terahertz instruments for space developed by various agencies

Mission/Instrument	Agency	Frequency of operation	
Salyut 6	Soviet Union	200 GHz	1978
COBE	NASA	30 GHz–3 THz	1989
SWAS	NASA	490–550 GHz	1998
Herschel, Planck	ESA	0.48–1.9 THz	2008
Millimetron	Russia	150 GHz–1.5 THz	2015?

diameter antenna with 50 dB Gain, 10 mW transmitted power with a bandwidth of 50 GHz gives us data rates as high as 300 Gbps with a 50 channel bandwidth.

Table 6 lists the roadmap of the terahertz instruments space exploration (Siegel 2010). The characteristics of terahertz radiation are in the range of 1.2–12.4 meV, which is equal to a blackbody temperature of 14–140 K, corresponding to the vibrational motion of many atoms and molecules. These characteristics form the basis of submillimeter astronomy and have been the primary stimulus to the research and development of terahertz instruments. Temperature and water vapor description, chemical signatures representative of our atmosphere provide us with critical information, which in turn is deciphered by terahertz spectroscopy. Thus, in order to draw these crucial observations, an unhindered environment is necessary. This promotes the use of these instruments in performing direct detection or heterodyne-based spectroscopy. From Salyut 6 space station's, first terahertz observation to the highly sophisticated Herschel Space Observatory which has pushed the boundaries on receiver and source technologies there has been an overwhelming effort for further improvements in terahertz instruments as researchers seek to tap information from space and the Earth's atmosphere (Siegel 2007).

Figure 23 shows the various types of terahertz instruments according to the operating frequency and sensitivity (Gaidis 2000). Concentrations in parts per billion and spectral resolution of the order 10^6 (or higher) often need to be handled in astronomical observations. Thus high sensitivity in space applications is paramount; as a result heterodyne devices are most popular in such applications.

Heterodyne devices are coherent receivers/detectors. Figure 24 is a block diagram of a terahertz heterodyne receiver. A signal with frequency f_s is picked by an antenna which is then mixed with a sine wave generated by a local oscillator. This combination in the mixer generates beats with a frequency f_{IF} (intermediate frequency) which is then demodulated. The intermediate frequency signal is processed and preservation of the amplitude and phase information is ensured (Chattopadhyay 2010). The f_{IF} is very small compared to the detected signal f_s giving high resolution. Terahertz instruments mostly employ SIS mixers or hot electron bolometers (HEB). But Schottky mixers provide an advantage with respect to the constraints on the payload and power; they also operate at room temperatures (Gaidis et al. 2000).

Schottky mixers, as frequency increases are constrained by skin effect which restricts the flow of current along the edge of the substrate; this increases the resistance in series for the diode (Bruston et al. 2000).

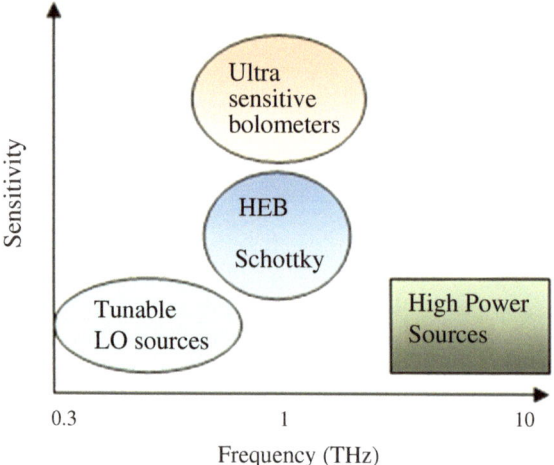

Fig. 23 Terahertz instruments corresponding to frequency of operation and sensitivity

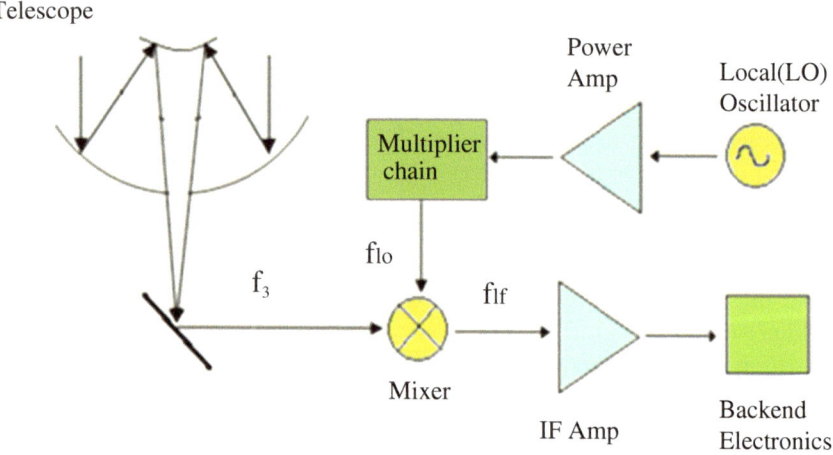

Fig. 24 Terahertz heterodyne receiver schematic

Thomas et al. (2010) first demonstrated a radiometer on chip (ROC) architecture which combines GaAs MMIC Schottky diodes and Si micromachining methods. This integration results in a significant reduction in the weight of the front end components which include the local oscillator, the basic mixers, amplifiers, and biasing circuits.

The current research trends in THz technology for space applications are focussed on the intersatellite links, telemetry of aircrafts during blackout, etc. The high data rate, secure communication, and high capacity are some of the advantages of utilizing THz technology in the space applications. The huge atmospheric loss encountered by the THz waves can be mitigated by implementing THz to Ka band conversion for satellite to earth link. THz waves can be used for the signal transmission from a GEO satellite to LEO satellite as well as among GEO satellites. The signal will be transmitted as Ka band waves from satellites to earth stations (Han et al. 2015).

The most challenging issue encountered by the THz communication is the requirement of a high power amplifier. The various high power amplifiers such as heterojunction transistors, nanowire FETs can provide high power for THz communication. Further, power combining using antenna array is a reliable technique for high power realization. Another major concern in THz systems is the design of THz antennas. High gain antennas with suitable beamwidth for tracking and acquisition are preferred in THz communication. The channel models and coding schemes need to be modified for the propagation of THz waves.

5.1 Deep Space Network

Since the start of the space exploration program, transmitting and receiving data back and forth from the spacecrafts and Earth has been a major challenge.

Communicating with the space probes on expeditions to various parts of our solar systems is far more difficult as compared to satellite communications. Communication signals decrease with the square of the distance between the receiver and the transmitter. The distances encountered in deep space communications are very large as compared to the distances between satellites and the ground stations. A 10 Gbps link with a GEO satellite at 36,000 kms would only realize 1 bps from Neptune which is at a distances of the order of 4 billion kms. Thus, an important characteristic required from deep space links (Fig. 25) to achieve high data rates is high out power and low divergence. High directivity requires accurate tracking and pointing. This pointing is subject to not only the antenna divergences, but also the attitude changes of the spacecraft.

Most of the deep space network consists of ground-based antenna clusters. NASA's deep space network consists of a cluster of couple of 34 m wave guide antennas and one 70 m Cassergrain antenna, each at Goldstone, California, Madrid, and Canberra. These facilitate communications with the deep space missions. In the pursuit for high data rates, there is a need to optimize all the system components involved. This can be achieved by manufacturing large apertures for the ground systems, development of novel low noise receivers, high power uplinks, and improved downlink capabilities. A phased antenna array which incorporates the advancements in MMICs and MEMS, cryogenic techniques to fabricate cheaper, reliable reflector antennas, receivers, and detectors is an attractive option to

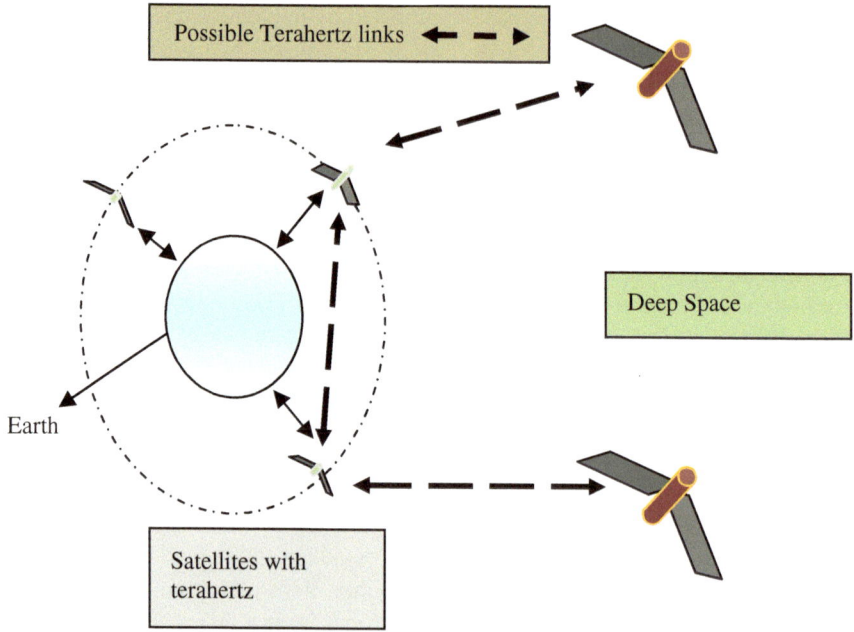

Fig. 25 Indirect links in a deep space network

replicate the characteristics of a 70 m antenna in a much more feasible way. Gatti (2008) reported the development of a phased antenna array formed by two 6 m antennas and one 12 m antenna.

Indirect links send and receive signals from satellites outside the Earth's atmosphere. Implementation of terahertz links would be more efficient with indirect links, this eliminates atmospheric attenuation. Adverse space environments imply that the spacecrafts need to be stable thermally. In order to ease the payload requirements, light weight, integrated telescopes are imperative. Antennas for the deep space probes are limited by the size and mass of the launching vehicle. Mission planners envisaged the requirement of deployable antennas. Deployed antennas provide the flexibility of compactness during the launch; they can later be deployed while ensuring restricted usage of the spacecraft resources (Schulze et al. 2007). In case of the failure to deploy, researchers have come up with a hybrid deployable antenna (HDA) system. A HDA consists of a fixed parabolic antenna along with a deployable annulus. This increases the antenna aperture significantly when deployed. The Russian space observatory mission Millimetron, scheduled to launch in 2015 has a 12 m deployable antenna and a 3.5 m solid mirror with frequency coverage of 150 GHz to 2 THz (Wild et al. 2008).

6 Conclusions

Terahertz technology is gaining momentum in space communication and security, because of the non-ionizing nature and penetration capability of terahertz radiation. This brief explores the terahertz antenna technology, toward implementation of compact, consistent, and cheap terahertz sources, and high sensitivity terahertz detectors, and the gain of the systems has become much more convenient. The advances in new materials and nano-structures, such as graphene will be helpful in miniaturization antenna technology while maintaining the desired output levels. Further PBG structures, metamaterials have helped in optimizing the existing technologies to give improved directivities and gains. Terahertz antenna characterization of bandwidth, impedance, polarization, etc., has not yet been methodically structured and it continues to be a major research challenge. This brief also explores the advances of terahertz technology in space applications worldwide, along with possibilities of use of this technology in deep space networks.

References

Akyildiz, F., J.M. Jornet, and C. Han. 2014. Terahertz band: next frontier for wireless communications. *Physical Communication* 12: 16–32.

Bowen, J.W., S. Hadjiloucas, B.M. Towlson, L.S. Karatzas, S.T.G. Wootton, N.J. Cronin, S.R. Davies, C.E. McIntosh, J.M. Chamberlain, R.E. Miles and R.D. Pollard. 2006. Micromachined waveguide antennas for 1.6 THz. *Electronics Letters* 42: 842–843.

Bruston, J., E. Schlecht, A. Maestrini, F. Maiwald, and S.C. Martin. 2000. Development of 200 GHz to 2.7 THz multiplier chains for submillimeter-wave heterodyne receivers. *Proceedings of SPIE* 4013: 285–295.

Carrasco, E., and J. Perruisseau-Carrier. 2013. Reflectarray antenna at terahertz using graphene. *IEEE Antennas and Wireless Propagation Letters* 12: 253–257.

Chen Wang, Xu, W. Sheng Zhao, J. Hu, and W. Yan Yin. 2015. Reconfigurable terahertz leaky-wave antenna using graphene-based high-impedance surface. *IEEE Transactions on Nanotechnology* 14: 62–69.

Chattopadhyay, G. 2010. Terahertz antennas and systems for space borne platforms. *Proceedings of 4th European Conference on Antennas and Propagation (EUCAP)*, pp. 1–7.

Danana, B., B. Choudhury, and R.M. Jha. 2014. Design of high gain microstrip antenna for THz wireless communication. *International Journal of Advanced Research in Electrical, Electronics and Instrumentation Engineering* 3(5): 711–716.

de Maagt, P., B.A. Conchillo, L. Minelli, I. Ederra, R. Gonzalo, and A. Reynolds. 2003. Photonic bandgap antennas and components for microwave and sub-millimeter wave applications. *IEEE Transactions on Antennas and Propagation* 51(10): 2667–2677.

Dragoman, M., A. A. Muller, D. Dragoman, F. Coccetti and R. Plana. 2010. Terahertz antenna based on graphene. *Journal of Applied Physics* 107: 104313(1)–104313(3).

Fitch, M.J., and R. Osiander. 2004. Terahertz waves for communications and sensing. *Johns Hopkins APL Technical Digest* 25(4): 348–355.

Gaidis, M. C. 2000. Space-based applications of far-infrared systems. *Proceedings of 8th International Conference on Terahertz Electronics*, pp. 125–128.

Gaidis, M.C., H.M. Pickett, C.D. Smith, S.C. Martin, R.P. Smith, and P.H. Siegel. 2000. A 2.5 THz receiver front end for space borne applications. *IEEE Transactions on Microwave Theory and Technology* 48(4): 733–739.

Gatti, M. 2008. A phased array antenna for deep space communications. *IEEE Aerospace Conference* 8 p.

Goldin, A., J. J. Bock, C. Hunt, A. Lange, H. G. LeDuc, A. Vayonakis,and J. Zmuidzinas. 2002. SAMBA: Superconducting antenna-coupled multi-frequency bolometric array. *Proceedings of far-IR, submm, and mm Detector Workshop,* vol. NASA CP/211408. Washington DC: NASA.

Grade, J., P. Haydon, and D. van der Weide. 2007. Electronic terahertz antennas and probes for spectroscopic detection and diagnostics. *Proceedings of the IEEE* 95(8): 1583–1591.

Han, H., J. Yuan, and J. Tong. 2015. Design of THz space application system. *Journal of Computer and Communications* 3: 61–65.

Hanson, G.W., and P. de Maagt. 2007. Guest editorial for the special issue on optical and THz antenna technology. *IEEE Transactions on Antennas and Propagation* 55(11): 2942–2943.

Huang, K.C., and Z. Wang. 2011. Terahertz terabit wireless communication. *IEEE Microwave Magazine* 12(4): 108–116.

Huang, Y., N. Khibiani, Y. Shen, and D. Li. 2011. Terahertz photoconductive antenna efficiency. *Proceedings of International Workshop on Antenna Technology (iWAT)*, pp. 152–156.

Jih, C., S. U. Hwu, and K. B. deSilva. 2013. Terahertz (THz) wireless systems for space applications. *Proceedings of IEEE Sensors Applications Symposium*, 5p.

Jornet, J.M., and I.F. Akyildiz. 2013. Graphene-based plasmonic nano-antenna for terahertz band communication in nano-networks. *IEEE Journal on Selected Areas in Communications/Supplement* 31(12): 685–694.

Khiabani, N., Y. Huang, Y. Shen, S. Boyes, and Q. Xu. 2013. A novel simulation method for THz photoconductive antenna characterization. *Proceedings of 7th European Conference on Antennas and Propagation (EUCAP)*, pp. 751–754.

Li, D., and Y. Huang. 2006. Comparison of terahertz antennas. *Proceedings of First European Conference on Antennas and Propagation*, pp. 1–5.

Li, D., Y. Huang, Y, Shen, A. Boland-Thoms, and A. Vickers. 2010. Development of a THz photoconductive horn antenna. *Proceedings of Fourth European Conference on Antennas and Propagation*, pp. 1–5.

Liu, Y., S. Li, S. Zhu, and X. Lv. 2010. New two-dimensional PBG structures for THz transmission line and antenna integrated design based on MEMS technology. *Proceedings of the IEEE*, pp. 1683–1686.

Llatser, I., C. Kremers, D. N. Chigrin, J. M. Jornet, M. C. Lemme, A. Cabellos Aparicio, and E. Alarcon. 2012. Characterization of graphene-based nano-antennas in the terahertz band. *Proceedings of 6th European Conference on Antennas and Propagation (EUCAP)*, pp. 194–198.

Llombart, N., C. Lee, M. Alonso-delPino, G. Chattopadhyay, C. Jung-Kubiak, L. Jofre, and I. Mehdi. 2013. Silicon micro-machined lens antenna for THz integrated heterodyne arrays. *IEEE Transactions on Terahertz Science and Technology* 3(5): 515–523.

Llombart, N. and G. Chattopadhyay. 2010. Extended hemispherical silicon lens excited by a leaky wave waveguide feed. *Proceedings of the European Conference on Antennas and Propagation*, Barcelona.

Llombart, N., A. Skalare, J. Gill, and P. H. Siegel. 2008. High efficiency submillimeter-wave imaging array. *Proceedings of the 33rd International Conference on Infrared, Millimeter, and Terahertz Waves*, Pasadena.

Moon, K., H. Han, and I. Park. 2005. Terahertz folded half wavelength dipole antenna for high output power. *Proceedings of International Topical Meeting on Microwave Photonics*, pp. 301–304.

Nagatsuma, T., H.J. Song, and Y. Kado. 2010. Challenges for ultrahigh-speed wireless communications using terahertz waves. *Terahertz Science and Technology* 3(2): 55–65.

Raisanen, A. V. 1998. Challenges in THz technology: applications, receivers and antenna testing. *Proceedings of 28th European Microwave Conference Amsterdam*, vol. 1, pp. 126–131.

Räisänen, A., J. Ala-Laurinaho, A. Karttunen, J. Mallat, A. Tamminen, and M. Vaaja. 2010. Measurements of high-gain antennas at THz frequencies. *Proceedings of the 4th European Conference on Antennas and Propagation (EUCAP)*, pp. 1–3.

Räisänen, A.V., J. Ala-Laurinaho, J. Häkli, A. Karttunen, T. Koskinen, A. Lönnqvist, J. Mallat, E. Noponen, A. Tamminen, M. Vaaja, and V. Viikari. 2007. How to test a high-gain antenna at THz frequencies?. *Proceedings of 19th International Conference on Applied Electromagnetics and Communications, ICECom*, pp. 1–3.

Sánchez-Escuderos, D., M. Ferrando-Bataller, A. Berenguer, and M. Baquero-Escudero. 2011. Design of low-loss waveguides and devices at THz frequencies using EBG structures. *Proceedings of the 5th European Conference on Antennas and Propagation (EUCAP)*, pp. 452–456.

Schneider, T., A. Wiatrek, S. Preussler, M. Grigat, and R.P. Braun. 2012. Link budget analysis for terahertz fixed wireless links. *IEEE Transactions on Terahertz Science and Technology* 2(2): 250–256.

Siegel, P. H. 2010. THz for space: the golden age. *Proceedings of IEEE MTT-S International Microwave Symposium Digest (MTT)*, pp. 816–819.

Siegel, P. H., P. de Maagt, and A. I. Zaghloul. 2006. Antennas for terahertz applications. *Proceedings of IEEE Antennas Propagation Symposium*, pp. 2383–2386.

Siegel, P.H. 2007. THz instruments for space. *IEEE Transactions on Antennas and Propagation* 55(11): 2957–2965.

Song, H.J., and T. Nagatsuma. 2011. Present and future of terahertz communications. *IEEE Transactions on Terahertz Science and Technology* 1(1): 256–263.

Syed, W.H., G. Fiorentino, D. Cavallo, M. Spirito, P.M. Sarro, and A. Neto. 2015. Design, fabrication, and measurements of a 0.3 THz on-chip double slot antenna enhanced by artificial dielectrics. *IEEE Transactions on Terahertz Science and Technology* 5(2): 288–298.

Thomas, B., C. Lee, A. Peralta, J. Gill, G. Chattopadhyay, E. Schlecht, R. Lin, and I. Mehdi. 2010. 600 GHz silicon-based integrated receiver using GaAs MMIC membrane planar Schottky diodes. *Proceedings of the 21st International Symposium on Space Terahertz Technology*.

Tonouchi, M. 2007. Cutting edge terahertz technology. *Nature Photonics* 1: 97–105.

Wild, W., A. Baryshev, T. de Graauw, N. Kardashev, S. Likhachev, G. Goltsman, and V. Koshelets. 2008. Instrumentation for Millimetron—a large space antenna for THz astronomy. *Proceedings of the 19th International Symposium on Space Terahertz Technology*, pp. 186–191.

Wu, K., Y. J. Cheng, T. Djerafi, and W. Hong. 2012. Substrate integrated millimetre wave and terahertz antenna technology. *Proceedings of the IEEE,* vol. 100, no. 7.

Xu, X., X. Zhang, Z. Zhou, T. Gao, Q. Zhang, Y. Lin, and L. Sun. 2013. Terahertz cassergrain reflector antenna. *Proceedings of the International Symposium on Antennas & Propagation (ISAP)*, vol. 2, pp. 969–971.

Yablonovitch, E. 1993. Photonic band-gap structures. *Journal of the optical society of America* 10 (2): 283–295.

Yong, L., L. Xin, and Y. Yong. 2006. Research and design of terahertz horn antenna based on MEMS technology. *Proceedings of 7th International Symposium on Antennas, Propagation and EM Theory*, pp. 1–3.

Appendix A
Prefixes

Multiplying factor	Prefix	Symbol
10^{12}	Tera	T
10^{9}	Giga	G
10^{6}	Mega	M
10^{3}	Kilo	k
10^{-3}	Milli	m
10^{-6}	Micro	μ
10^{-9}	Nano	n
10^{-12}	Pico	p
10^{-15}	Femto	f

© The Author(s) 2016
B. Choudhury et al., *Terahertz Antenna Technology for Space Applications*,
SpringerBriefs in Computational Electromagnetics,
DOI 10.1007/978-981-287-799-4

Appendix B
Physical Constants

- Permittivity of space $\varepsilon_0 = 8.854 * 10^{-12}$ F/m
- Permeability of space $\mu_0 = 4\pi * 10^{-7}$ H/m
- Velocity of light in free space $c = 2.99 * 10^8$ m/s
- Charge of electron $q = 1.602 * 10^{-19}$ C
- Mass of electron $m_e = 9.107 * 10^{-31}$ kg
- Boltzmann's constant $k_b = 1.38 * 10^{-23}$ J/K
- Planck's constant h = $6.626 * 10^{-34}$ J-s

© The Author(s) 2016
B. Choudhury et al., *Terahertz Antenna Technology for Space Applications*,
SpringerBriefs in Computational Electromagnetics,
DOI 10.1007/978-981-287-799-4

Appendix C
Maxwell's Equations

The general form of time varying Maxwell's equation:

$$\nabla \times \boldsymbol{E} = -\frac{\partial \boldsymbol{B}}{\partial t} - \boldsymbol{M}$$

$$\nabla \times \boldsymbol{H} = \frac{\partial \boldsymbol{D}}{\partial t} + \boldsymbol{J}$$

$$\nabla \cdot \boldsymbol{D} = \rho$$

$$\nabla \cdot \boldsymbol{B} = 0$$

$\boldsymbol{E}$ is the electric field vector (V/m)
$\boldsymbol{H}$ is the Magnetic field vector (A/m)
$\boldsymbol{D}$ is the Electric flux density vector (C/m^2)
$\boldsymbol{B}$ is the Magnetic flux density vector (Wb/m^2)
$\boldsymbol{J}$ is the Electric current density vector (A/m^2)
$\boldsymbol{M}$ is the Magnetic current density vector (V/m^2)
ρ is the electric charge density (C/m^3)

Constitutive Relations:

$$\boldsymbol{D} = \varepsilon \boldsymbol{E}$$

$$\boldsymbol{B} = \mu \boldsymbol{H}$$

ε, μ are the permittivity and permeability tensors respectively.

© The Author(s) 2016
B. Choudhury et al., *Terahertz Antenna Technology for Space Applications*,
SpringerBriefs in Computational Electromagnetics,
DOI 10.1007/978-981-287-799-4

Appendix D
Antenna Properties

1. Far field distances correspond to distances, where spherical wavefronts can be approximated to planar wavefronts. This approximation is restricted to the radiating aperture. Far field distance is defined as,

$$r \geq \frac{2 \times d^2}{\lambda}$$

   d is the maximum dimension of the antenna; λ is the wavelength of radiation.
2. Antenna near field can be divided into two regions named reactive near field and radiating near field. Generally the near field distance can be defined as

$$r \leq \frac{2 \times d^2}{\lambda}$$

3. Radiation pattern is a graphical representation of radiation properties such as electric field strength, radiation intensity, and polarization of an antenna with respect to spatial coordinates.
4. Half power beamwidth is the angular separation at which the radiation intensity becomes half of its maximum value.
5. Radiation power density is given as

$$W_{rad} = \frac{1}{2} \mathrm{Re}[E \times H^*]$$

6. Radiation intensity is the power radiated from an antenna per unit solid angle

$$U = r^2 W_{rad}$$

7. Radiated power is given as

© The Author(s) 2016
B. Choudhury et al., *Terahertz Antenna Technology for Space Applications*,
SpringerBriefs in Computational Electromagnetics,
DOI 10.1007/978-981-287-799-4

$$P_{rad} = 4\pi U_{avg}$$

8. The ability of the antenna to focus power in a particular direction is a measure of its directivity. It is given as

$$D = \frac{U_{max}}{U_{avg}}$$

U_{max} is the maximum radiation intensity in the direction of the main beam. U_{avg} is the average radiation density in all directions.

9. Antenna gain is defined as the product of the directivity and the efficiency of the antenna. Antenna gain describes the directional properties taking into account the losses in the antenna.

$$G = \eta_{rad}D$$

10. Antenna efficiency considers the losses of the antenna at the input terminal and the structure such as reflection, conduction, and dielectric losses. Therefore, overall efficiency can be given as the product of reflection, conduction, and dielectric efficiencies.

$$\eta_t = \eta_r\eta_c\eta_d$$

11. Antenna radiation efficiency is the product of conduction and dielectric efficiency

$$\eta_{rad} = \eta_c\eta_d$$

12. Radiation efficiency is also given as the ratio of the radiated power to the input power.

$$\eta_{rad} = P_{rad}/P_{in}$$

13. Maximum effective aperture area of an antenna at an operating wavelength λ is given as

$$A_{eff} = D\frac{\lambda^2}{4\pi}$$

14. The impedance bandwidth of an antenna is the range of frequencies for which the input impedance is within an acceptable level as that of the center frequency.

15. Antenna input impedance is the impedance as seen from a pair of input terminal which is equal to the ratio of voltage to current at that terminal.

16. Antennas which radiate equally in all directions are known as isotropic antennas. The transmitted power (P_t) and the transmitter antenna gain (G_t) are transmitter characteristics. Their product can be inferred as the power radiated by an isotropic antenna with power input $(P_t G_t)$. This product is known as the effective isotropic radiated power (EIRP).

$$EIRP = P_t G_t (W)$$

About the Book

This book explores the terahertz antenna technology toward implementation of compact, consistent, and cheap terahertz sources, as well as the high sensitivity terahertz detectors. The terahertz EM band provides a transition between the electronic and the photonic regions thus adopting important characteristics from these regimes. These characteristics, along with the progress in semiconductor technology, have enabled researchers to exploit hitherto unexplored domains including satellite communication, biomedical imaging, and security systems. The advances in new materials and nanostructures such as graphene will be helpful in miniaturization of antenna technology while simultaneously maintaining the desired output levels. Terahertz antenna characterization of bandwidth, impedance, polarization, etc. has not yet been methodically structured and it continues to be a major research challenge. This book addresses these issues besides including the advances of terahertz technology in space applications worldwide, along with possibilities of using this technology in deep space networks.

© The Author(s) 2016   45
B. Choudhury et al., *Terahertz Antenna Technology for Space Applications*,
SpringerBriefs in Computational Electromagnetics,
DOI 10.1007/978-981-287-799-4

Author Index

A

Akyildiz, F., 7, 19, 22
Ala-Laurinaho, J., 24
Alarcon, E., 19
Alonso-delPino, M., 17

B

Baryshev, A., 30
Bock, J.J., 17
Boland-Thoms, A., 11, 12
Boyes, S., 22
Braun, R.P., 6
Bruston, J., 27

C

Cabellos-Aparicio, A., 19
Carrasco, E., 19
Cavallo, D., 22
Chattopadhyay, G., 17, 18, 27
Chen, W.X., 21
Cheng, Y.J., 21
Chigrin, D.N., 19
Coccetti, F., 21
Conchillo, B.A., 16

D

Danana, B., 17
de Graauw, T., 30
deMaagt, P., 2, 4, 10
deSilva, K.B., 26
Djerafi, T., 21
Dragoman, D., 21
Dragoman, M., 20

E

Ederra, I., 16

F

Fiorentino, G., 22

Fitch, M.J., 3, 8, 9

G

Gaidis, M.C., 27
Gao, T., 16
Gatti, M., 30
Gill, J., 17, 28
Goldin, A., 17
Goltsman, G., 30
Gonzalo, R., 16
Grade, J., 5
Grigat, M., 6

H

Häkli, J., 24
Han, C., 7, 19, 22
Han, H., 13, 29
Hanson, G.W., 2, 4
Haydon, P., 5
Hong, W., 21
Hu, J., 21
Huang, K.C., 6
Huang, Y., 10–12, 22, 23
Hunt, C., 17
Hwu, S.U., 27

J

Jih, C., 27
Jofre, L., 18
Jornet, J.M., 7, 19, 22
Jung-Kubiak, C., 18

K

Kado, Y., 6
Kardashev, N., 30
Karttunen, A., 24
Khiabani, N., 22
Koshelets, V., 30
Koskinen, T., 24

© The Author(s) 2016
B. Choudhury et al., *Terahertz Antenna Technology for Space Applications*,
SpringerBriefs in Computational Electromagnetics,
DOI 10.1007/978-981-287-799-4

Kremers, C., 19

L
Lange, A., 17
LeDuc, H.G., 17
Lee, C., 18, 28
Lemme, M.C., 19
Li, D., 10, 11, 14
Li, S., 15
Likhachev, S., 30
Lin, R., 28
Lin, Y., 16
Llatser, I., 19
Llombart, N., 18
Lönnqvist, A., 24
Lv, X., 15

M
Maestrini, A., 27
Maiwald, F., 27
Mallat, J., 24
Martin, S.C., 27
Mehdi, I., 18, 28
Minelli, L., 16
Moon, K., 13
Muller, A.A., 21

N
Nagatsuma, T., 5, 6
Neto, A., 22
Noponen, E., 24

P
Park, I., 13
Peralta, A., 28
Perruisseau-Carrier, J., 19
Pickett, H.M., 27
Plana, R., 21
Preussler, S., 6

R
Räisänen, A.V., 24
Reynolds, A., 16

S
Sarro, P.M., 22
Schlecht, E., 27
Schneider, T., 6
Shen, Y., 11, 12, 22

Siegel, P.H., 10, 17, 18, 27
Skalare, A., 17
Smith, C.D., 27
Smith, R.P., 27
Song, H.J., 5, 6
Spirito, M., 22
Sun, L., 16
Syed, W.H., 22

T
Tamminen, A., 24
Thomas, B., 28
Tong, J., 29
Tonouchi, M., 4, 5, 9

V
Vaaja, M., 24
Vayonakis, A., 17
Vickers, A., 11
Viikari, V., 24

W
Wang, Z., 6
Weide, D., 5
Wiatrek, A., 6
Wild, W., 30
Wu, K., 21

X
Xin, L., 14
Xu, Q., 22
Xu, X., 16

Y
Yablonovitch, E., 16
Yan-Yin, W., 21
Yong, L., 14
Yong, Y., 14
Yuan, J., 29

Z
Zaghloul, A.I., 10
Zhang, Q., 16
Zhang, X., 16
Zhao, W.S., 21
Zhou, Z., 16
Zhu, S., 15
Zmuidzinas, J., 17

Subject Index

A
Arrays, 17
Artificial dielectric layers, 21

C
Carbon fiber reinforced polymers (CFRP), 14
Carbon nanotubes, 18
Cassergrain antenna, 29
Communications, 4
Cryogenic techniques, 29

D
Deep space network, 29

E
Electro optic effect, 8

F
Frequency domain spectroscopy system, 9

G
Graphene-based antennas, 19
Graphene metasurfaces, 21

H
High impedance surface, 21
Hybrid deployable antenna, 30

L
Leaky wave antennas, 21

M
MEMS-based antennas, 14

Microlens, 18
Micromachining, 2

N
Nanoantennas, 18
Nanowire FETs, 29

P
PBG waveguide, 15
Photocarrier density, 23
Photoconductive antennas, 10
Photo-mixing, 3
Photonic band gap-based antennas, 14

Q
Quantum cascade laser, 4

R
RF-hologram, 24

S
Submillimeter, 1
Substrate integrated antennas, 21

T
Terabit wireless local area networks, 6
Terahertz-TDS system, 8
Terahertz antenna testing, 24
Terahertz antennas, 10
Terahertz domain, 2
Terahertz photons, 3
Terahertz radiation, 1
Terahertz sources, 3

© The Author(s) 2016
B. Choudhury et al., *Terahertz Antenna Technology for Space Applications*,
SpringerBriefs in Computational Electromagnetics,
DOI 10.1007/978-981-287-799-4